Risk IT
Practitioner Guide

2nd Edition

ISACA.

About ISACA

For more than 50 years, ISACA® (*www.isaca.org*) has advanced the best talent, expertise and learning in technology. ISACA equips individuals with knowledge, credentials, education and community to progress their careers and transform their organizations, and enables enterprises to train and build quality teams. ISACA is a global professional association and learning organization that leverages the expertise of its 145,000 members who work in information security, governance, assurance, risk and privacy to drive innovation through technology. It has a presence in 188 countries, including more than 220 chapters worldwide.

Disclaimer

ISACA has designed and created the *Risk IT Practitioner Guide, 2nd Edition* (the "Work") primarily as an educational resource for professionals. ISACA makes no claim that use of any of the Work will assure a successful outcome. The Work should not be considered inclusive of all proper information, procedures and tests or exclusive of other information, procedures and tests that are reasonably directed to obtaining the same results. In determining the propriety of any specific information, procedure or test, professionals should apply their own professional judgment to the specific circumstances presented by the particular systems or information technology environment.

Reservation of Rights

ISACA

1700 E. Golf Road, Suite 400
Schaumburg, IL 60173, USA
Phone: +1.847.660.5505
Fax: +1.847.253.1755
Contact us: https://support.isaca.org
Website: www.isaca.org

Participate in the ISACA Online Forums: https://engage.isaca.org/onlineforums

Twitter: http://twitter.com/ISACANews
LinkedIn: www.linkedin.com/company/isaca
Facebook: www.facebook.com/ISACAGlobal
Instagram: www.instagram.com/isacanews/

Risk IT Practitioner Guide, 2nd Edition
ISBN 978-1-60420-822-1

Acknowledgments

ISACA would like to recognize

Lead Developer
Lisa Young, CISA, CISM, CISSP, Axio, USA

Developer
Dirk Steuperaert, CISA, CRISC, CGEIT, IT In Balance, Belgium

Expert Reviewers
Luis Alberto Capua, CRISC, CISM, Argentina
Tom Conkle, CISSP, Optic Cyber Solutions, USA
Andrew Foo, CISA, CRISC, CISM, CGEIT, CBCP, CCSK, CISSP, PMP, Dulwich College International, China
Sandra Fonseca, Ph.D., CISA, CRISC, CISM, CICA, Northcentral University, USA
Yalcin Gerek, CISA, CRISC, CGEIT, COBIT 5 Trainer, DASA DevOps Coach, ISO 20000LI, ISO 27001LA,
    ITIL Expert, PRINCE2, Resilia Practitioner, TAC, Turkey
Ahmad M. El Ghazouly, Ph.D., CISA, CRISC, CISM, PMI-ACP, AMBCI, BISL, PBA, PMP, PMI-RMP, TOGAF,
    PGESCo, Egypt
Demetri Gittens, CISA, CRISC, Central Bank of Trinidad and Tobago, Trinidad & Tobago
Ken Hendrie, CISA, CRISC, CISM, CGEIT, ISO27001 LI, ITIL, PRINCE2, IRAP, Cyconsol, Australia
John Hoffoss, CISA, CISSP, CGIH, CliftonLarsonAllen, USA
Mike Hughes, CISA, CGEIT, CRISC, MIoD, Prism RA, United Kingdom
John E. Jasinski, CISA, CRISC, CISM, CGEIT, CSX, COBIT 5 Assessor, COBIT and ITIL Accredited Instructor,
    AWS Practitioner, CCSK, Certified Scrum Master and Product Owner, ISO 20000, IT4IT, ITIL Expert, Lean IT,
    MOF, ServiceNow and RSA Archer Certified System Administrator, Six Sigma Blackbelt, TOGAF, USA
Jack Jones, CISA, CISM, CRISC, CISSP, RiskLens, USA
Linda Kostic, CISA, CISSP, CPA, Doctor of IT - Cybersecurity & Information Assurance, PRMIA Complete Course
    in Risk Management at George Washington University (GWU), Citi, USA
Jerry M. Kathingo, CRISC, CISM, Hatari Security, Kenya
Kamal Khan, CISA, CISSP, CITP, MBCS, United Kingdom
Shruti S. Kulkarni, CISA, CRISC, CCSK, CISSP, ITIL v3, Interpublic Group, United Kingdom
Jim Lipkis, Monaco Risk Analytics Inc., USA
Tony Martin-Vegue, CISM, CISSP, Netflix
Andre Pitkowski, CRISC, CGEIT, COBIT 5 Assessor, APIT Consultoria de Informatica Ltda, Brazil
Eduardo Oscar Ritegno, CISA, CRISC, Banco Nación, Argentina
Gurvinder Pal Singh, CISA, CRISC, CISM, Qantas Airways, Australia
Katsumi Sakagawa, CISA, CRISC, Japan
Darron Sun, CISA, CRISC, CISSP, CMA, CPA (Australia), CRMA, FIPA, Hong Kong Housing Society, China
Peter C. Tessin, CISA, CRISC, CISM, CGEIT, Discover Financial Services, USA
Alok Tuteja, CRISC, CGEIT, CIA, CISSP, BRS Ventures, United Arab Emirates
Ashish Vashishtha, CISA, CRISC, CISM, CIPT, CISSP, AWS Certified Cloud Practitioner, HITRUST CSF
    Practitioner, PROSCI Change Practitioner, AdventHealth, USA
Greet Volders, CGEIT, Voquals N.V., Belgium
Jonathan Waldo, CISA, CRISC, ITIL 4 Foundation, CH Robinson, USA
Larry G. Wlosinski, CISA, CRISC, CISM, CAP, CBCP, CCSP, CDP, CIPM, CISSP, ITIL v3, PMP, Coalfire-Federal, USA
Prometheus Yang, CISA, CISM, CRISC, CFE, Standard Chartered Bank, Hong Kong
Dušan Žikić, CISA, CRISC, CISM, CSX-P, Cybersecurity Audit, Cybersecurity Fundamentals, COBIT 5
    Foundation, COBIT 2019 Foundation, COBIT 2019 Design and Implementation, ITIL (2011) Foundation,
    ITIL 4 Foundation, IBM Data Science, NIS Gazprom Neft, Serbia

Acknowledgments (cont.)

Risk IT Task Force

Steven Babb, CRISC, CGEIT, ITIL, MUFG Investor Services, United Kingdom
Urs Fischer, CISA, CRISC, CPA (Swiss), UBS Business Solutions AG, Switzerland
Jack Freund, Ph.D., CISA, CRISC, CISM, CISSP, RiskLens, USA
Apolonio Garcia, CRISC, Open FAIR, HealthGuard, USA
Jimmy Heschl, CISA, CISM, CGEIT, Red Bull, Austria
Gladys Rouissi, CISM, CRISC, ANC Wealth, Australia
James C. Samans, CISA, CRISC, CISM, CBCP, CISSP-ISSEP, CPP, PMP, American Institutes for Research, USA
Ekta Singh-Bushell, CISA, CGEIT, CISSP, CPA, Datatec, USA
Dirk Steuperaert, CISA, CRISC, CGEIT, IT In Balance, Belgium
Evan Wheeler, CRISC, IASO, Edelman Financial Engines, USA

Board of Directors

Tracey Dedrick, Chair, Former Chief Risk Officer, Hudson City Bancorp, USA
Rolf von Roessing, Vice-Chair, CISA, CISM, CGEIT, CDPSE, CISSP, FBCI, Partner, FORFA Consulting AG, Switzerland
Gabriela Hernandez-Cardoso, Independent Board Member, Mexico
Pam Nigro, CISA, CRISC, CGEIT, CRMA, Vice President–Information Technology, Security Officer, Home Access Health, USA
Maureen O'Connell, Board Chair, Acacia Research (NASDAQ), Former Chief Financial Officer and Chief Administration Officer, Scholastic, Inc., USA
David Samuelson, Chief Executive Officer, ISACA, USA
Gerrard Schmid, President and Chief Executive Officer, Diebold Nixdorf, USA
Gregory Touhill, CISM, CISSP, President, AppGate Federal Group, USA
Asaf Weisberg, CISA, CRISC, CISM, CGEIT, Chief Executive Officer, introSight Ltd., Israel
Anna Yip, Chief Executive Officer, SmarTone Telecommunications Limited, Hong Kong
Brennan P. Baybeck, CISA, CRISC, CISM, CISSP, ISACA Board Chair, 2019-2020, Vice President and Chief Information Security Officer for Customer Services, Oracle Corporation, USA
Rob Clyde, CISM, ISACA Board Chair, 2018-2019, Independent Director, Titus, and Executive Chair, White Cloud Security, USA
Chris K. Dimitriadis, Ph.D., CISA, CRISC, CISM, ISACA Board Chair, 2015-2017, Group Chief Executive Officer, INTRALOT, Greece

TABLE OF CONTENTS

Page intentionally left blank

LIST OF FIGURES

Page intentionally left blank

Chapter 1
Introduction to the *Risk IT Practitioner Guide*

1.1 Overview of Risk IT

The *Risk IT Practitioner Guide* is about information and technology (I&T) risk—in other words, business or mission risk related to the use of, or dependence on, information and communications technology (ICT),[1] operational technology (OT),[2] network or Internet of Things (IOT),[3] electronic data, and digital or electronic communications. The connection to business or mission originates in the principles on whichthe *Risk IT Framework, 2nd Edition* published separately as a companion document to this guide, is built, that is, effective enterprise governance and management of all types of information- and technology-related risk.

The I&T designation as used in this document includes, but is not limited to, information security, information assurance and cybersecurity as subtypes of I&T risk.

The *Risk IT Practitioner Guide* (this publication) contains practical and sometimes more detailed guidance on how to accomplish some of the activities described in its companion publication, the *Risk IT Framework*. The *Risk IT Framework* describes the tasks and activities, or more simply, the "what to do" for the management of I&T risk. In this publication, references are made to terms and concepts that are introduced in the *Risk IT Framework* such as "risk taxonomy," "risk appetite," "risk identification," "risk analysis" and "key risk indicators" (KRIs). A reader of this document may want to become acquainted with the *Risk IT Framework* definitions and appendix B of this publication before applying the tasks and activities to their enterprise.

Risk management is a business capability or, simply, a capability an enterprise operates that describes a collective set of activities that are applied to achieve a specific outcome. The risk management capability underpins all business processes to ensure the enterprise continues to create and deliver value. Risk management activities require people, processes and technology and may be automated in software and systems as well as overseen by the highest levels of the organization. Risk management works best when integrated with the regular workflow of the staff and management rather than as an add-on activity.

The benefits of effective I&T risk management include the following:

- Better oversight of organizational assets
- Reduced or minimized losses
- More cost-effective use of resources
- Identification of threats, vulnerabilities and potential impacts on a proactive basis
- Prioritization of risk response efforts to match organizational goals and priorities
- A more holistic basis for and approach to legal and regulatory compliance
- Increased likelihood of project success
- Improved performance in the enterprise, leading to greater stakeholder confidence
- Creation of a risk-aware culture with less reliance on only risk specialists
- Improved incident and business continuity management
- Considerations for efficient control design with better monitoring and reporting
- Improved decision making as a result of expanded access to accurate, timely information
- An increased ability to meet business objectives and create value

[1] NIST, "Information and Communications Technology (ICT)," Computer Security Resource Center, *https://csrc.nist.gov/glossary/term/information_and_communications_technology*
[2] NIST, "Operational Technology (OT)," Computer Security Resource Center, *https://csrc.nist.gov/glossary/term/operational_technology*
[3] Voas, Jeffrey; "Network of 'Things'," 800-183, NIST, July 2016, *https://csrc.nist.gov/publications/detail/sp/800-183/final*

For readers looking for an overall structure in which to think about risk management, refer to the *Risk IT Framework* (published separately).

For readers looking for guidance on the governance of risk management, chapters 1, 2, 3 and 6 of the *Risk IT Practitioner Guide* are most applicable.

For readers looking for guidance on the management or considerations for implementation of risk management activities, chapters 1, 2, 4, 5 and 6 of the *Risk IT Practitioner Guide* are most applicable.

For those readers who are more familiar with the COBIT® 2019 processes and in keeping with the ISACA principles that governance and management are distinct sets of activities, this document is divided accordingly into sections covering each distinct area of governance and management. See **figure 1.1**.

1.1.1 The Language of Risk

Meaningful I&T risk assessments and risk-based decisions require I&T risk to be expressed in unambiguous and clear business or mission-relevant terms related to concerns such as finance, revenue or ability to meet desired strategic outcomes. Effective risk management requires mutual understanding between I&T and the rest of the business over which risk needs to be managed and why. All stakeholders must have the ability to understand and express how adverse events, also known as realized risk or incidents, may affect business or mission objectives. This means that there is a shared understanding that:

- I&T-related failures, compromises, mistakes or events can impact enterprise objectives and result in the loss of direct costs (e.g., financial) or indirect information (e.g., customer-sensitive data), resulting in reputational damage.
- Losses to the enterprise from I&T-related events can affect the ability for an organization to deliver its key services and products. This is true even when the enterprise relies upon I&T suppliers to provide goods and resources that are integral to its strategic objectives.

Communicating about risk requires that the terms used in the enterprise to express and describe risk must have a commonly understood meaning. A risk **taxonomy** provides a scheme for classifying sources and categories of risk. The path from a cyberthreat or area of concern to a risk requires that the statement of risk be decomposed into components that are actionable. A risk taxonomy provides a common language for discussing and communicating risk to stakeholders. See Appendix A: Risk Resources for informative references on resources related to risk, risk taxonomies and risk management.

The key concepts of risk are discussed in various contexts throughout this publication. It is common for people who lack strong understanding of risk terms to use them interchangeably, but doing so can create confusion, impede successful risk management and cast doubt on professional competence. For example, when the terms "threat," "vulnerability," "issue" and "risk" are used interchangeably or inconsistently it is not always clear what is being communicated. The risk practitioner should ensure that he/she spends enough time studying the language to gain a basic, reliable understanding of the different terms and how they relate to one another. For further discussion on risk terminology, refer to chapter 1 of the *Risk IT Framework* (published separately).

Risk can be discussed in quantitative (using numbers) or qualitative (using descriptive words) terms, and the specific definitions of risk sometimes vary from source to source. The fundamental nature of risk is uncertainty. This means that the uncertainty of risk being realized and resulting in loss or damage may or may not occur and is generally discussed related to risk using terms like "probability," "likelihood," "volatility" and "frequency." Another part of the uncertainty is what the loss or damage would mean for the organization if that risk materializes and is generally discussed using terms such as "impact," "magnitude" or "consequence." Early attempts to define risk observed that the probability of something happening was a combination of two things: whether something with the potential for harm occurred (e.g., denial of service [DoS] attacks, emails mistakenly sent to the wrong recipient, dangerous weather events) and whether the target of the event was susceptible to the attempt (vulnerability). Much of the time an enterprise has little control over the threats, vulnerabilities or other conditions in the environment in which they operate. However, the enterprise does have direct control over how risk is identified, assessed or analyzed, and managed.

As the practice of risk management has developed, risk practitioners have begun to distinguish between the conditions (risk factors) and the extent to which those factors affect the value-creation activities of the organization (impact). It is now common to distinguish between different types of threats, to evaluate them on the basis of specific organizational I&T assets against which they may be directed, and to assess those assets in terms of their individual weaknesses (vulnerabilities) that might be exploited to impact the business or mission.

As the practice of risk management has developed, risk practitioners have begun to distinguish between the conditions (risk factors) and the extent to which those factors affect the value-creation activities of the organization (impact).

The *Risk IT Framework* and the *Risk IT Practitioner Guide* are designed to assist in developing, implementing or enhancing the practice of risk management by:

- Connecting the business context with the specific I&T assets
- Shifting the focus to activities over which the enterprise has significant control, such as actively directing and managing risk, while minimizing the focus on the conditions over which an enterprise has little control (threat actors)
- Increasing the focus on using a common risk language that correctly labels the items that have to be managed well to create value

When viewed from the perspective of how I&T assets are used within the organization, the assets have value because of the business or mission purpose they serve. An I&T asset, by itself, may be easily replaceable, such as server hardware, or highly susceptible to vulnerabilities, such as software. However, without the business or mission context, it is not possible to fully understand the criticality of I&T assets. This shift from risk assessment (generally qualitative) to risk analysis (usually quantitative) makes it possible to more clearly communicate impact in terms of lost productivity and other specific measures of value, which is useful for many reasons:

1. Using a quantitative value of loss (or gain), based in the language of currency, time, productivity units or other measurable values, is easier to communicate to everyone.

2. Defining every consequence that is or is not acceptable in dozens or more different areas of business or mission functions, using only qualitative labels (low, high, catastrophic) and differing scales of measurement, does not allow for confident or justified decision-making about risk.

3. Quantifying the potential loss or other negative impact associated with risk provides a basis for deciding how to respond to risk that is out of tolerance with acceptable levels and where a traditional controls-based mitigation approach may not be applicable.

4. All I&T assets are not equal, and the budget needed to respond to risk should be commensurate with the value or criticality of the asset in the delivery of the business and mission objectives.

In enterprises wishing to enhance their risk management practices, the *Risk IT Practitioner Guide* can provide a solution accelerator, not in a prescriptive manner, but as a solid platform upon which an improved risk management practice can be built. The *Risk IT Practitioner Guide* can be used to assist with setting up an I&T risk management structure in the enterprise and to enhance existing I&T risk management practices.[4]

---

[4] This guide does not claim to be complete or comprehensive. In addition to the techniques and practices described here, other viable solutions and techniques exist and may be applied for managing I&T risk.

RISK IT PRACTITIONER GUIDE, 2ND EDITION

The Risk IT framework is described in full detail in the *Risk IT Framework* publication. For ease of reference, **figure 1.1** contains a graphic overview of the Risk IT framework and its components and how the principles align with COBIT objectives EDM03 *Ensured risk optimization* and APO12 *Managed risk*.

Figure 1.1—Risk IT Framework

Connect to enterprise objectives | Align with ERM | Balance cost/benefit of I&T-related risk | Promote ethical and open communication | Establish tone at the top and accountability | Use a consistent approach aligned to strategy

Risk Management Principles

Aligns to EDM03 Ensured Risk Optimization

Risk Governance
Ensure that the enterprise's risk appetite and tolerance are understood, articulated and communicated, and that risk to enterprise value related to the use of I&T is identified and managed.

Direct risk management.
Direct the establishment of risk management practices to provide reasonable assurance that I&T risk management practices are appropriate and that actual I&T risk does not exceed the board's risk appetite.

Monitor risk management.
Monitor the key goals and metrics of the risk management processes. Determine how deviations or problems will be identified, tracked and reported for remediation.

Evaluate risk management.
Continually examine and evaluate the effect of risk on the current and futureuse of I&T in the enterprise. Consider whether the enterprise's risk appetite is appropriate and ensure that risk to enterprise value related to the use of I&T is identified and managed.

Aligns to APO12 Managed Risk

Risk Management
Continually identify, assess and reduce I&T-related risk within tolerance levels set by enterprise executive management.

Collect data. Identify and collect relevant data to enable effective I&T-related risk identification, analysis and reporting.

Analyze risk. Develop a substantiated view on actual I&T risk, in support of risk decisions.

Maintain a risk profile. Maintain an inventory of known risk and risk attributes, including expected frequency, potential impact and responses. Document related resources, capabilities and current control activities related to risk items.

Articulate risk. Communicate information on the current state of I&T-related exposures and opportunities in a timely manner to all required stakeholders for appropriate response.

Define a risk management action portfolio. Manage opportunities to reduce risk to an acceptable level as a portfolio.

Respond to risk. Respond in a timely manner to materialized risk events with effective measures to limit the magnitude of loss.

Source: ISACA, *The Risk IT Framework, 2nd Edition*, USA, 2020, figure 3.1, *https://www.isaca.org/bookstore/bookstore-risk-digital/ritf2*

Additional guidance is available in the ISACA white paper *Getting Started with Risk Management*, the *Risk IT Framework* publication and references to COBIT® where appropriate.

Chapter 2
Setting the Context and Scoping Risk Management Activities

2.1 Setting the Context for Risk Management

Positioning risk in the context of the mission, strategy and objectives of the enterprise is the first step in making sure that risk management activities add value to the overall risk management process for the enterprise. This is known as setting the context for risk management. Pairing a risk-based approach with a strategic view of the enterprise enables communication and clarification of which uncertainties, or risk, have the highest potential to prevent the enterprise from meeting its intended targets, objectives and mission. For those practitioners who are more familiar with COBIT 2019, the activities in this chapter are related to APO12 *Managed risk.*[5]

Establishing the criteria against the identified risk that will be evaluated is also an important part of the overall risk management process. The development of risk appetite and risk tolerances can assist in quickly evaluating and understanding which risk is in alignment with management's objectives for risk-taking and which risk needs further analysis or investigation to make that determination.

Managing I&T risk of the enterprise starts with defining the risk universe. A risk universe describes the overall (risk) environment (i.e., defines the boundaries of risk management activities) and provides a structure for managing I&T risk. The selection of items included in the risk activities is generally based upon understanding the full risk universe and then selecting the specific part of the enterprise to which the risk activities will be applied. This is often called risk scoping. The risk universe:

- Considers the overall mission and the enterprise objectives, business processes and dependencies throughout the enterprise. Identification of I&T dependencies will aid in understanding the risk that cuts across different functions and operations of the organization.

- Describes the I&T components, processes, assets and infrastructure that support the business and mission objectives

- Describes risk in complete and comprehensive language so it can be viewed from an end-to-end business or mission perspective

- Considers the full value chain of the enterprise to include subsidiaries, business units, clients, suppliers and service providers. These designations may be candidates to consider in scoping the risk activities.

- Includes a logical and workable segmentation of the overall risk landscape in which the enterprise operates (e.g., organizational units and subunits; business processes or services; geographic locations; technology types, such as internal IT function vs. cloud components; and other areas where there may be an opportunity to align differing views across the enterprise).

- Aligns the strategic planning of the organization with the identification of the risk types that would have the greatest impact on meeting the business strategy and objectives

- Is also influenced by the business climate, or geopolitical environment, in which the enterprise operates

2.1.1 Scoping I&T Risk Management Activities

Risk management requires that the scope of the risk landscape, or risk universe, is set and the criteria against which the identified risk will be assessed or evaluated are defined. Risk management starts with understanding the organization, but the risk practitioner should bear in mind that the organization is heavily influenced by the environment, or context, in which it operates. This is particularly true of organizations that operate in a single sector

---

[5] See ISACA, *COBIT® 2019 Framework: Governance and Management Objectives*, USA, 2018, *https://www.isaca.org/bookstore/bookstore-cobit_19-print/cb19fgm.*

of the economy, such as financial services, manufacturing or healthcare, because there is a heavy dependency on many of the same supply chains.

The scope should be determined within the context of the organization's objectives. Setting the context will help to identify where the scope of the initial risk assessment (e.g., one business function, such as accounting) fits within the context of the overall enterprise. Defining an initial, preliminary scope for risk management activities can be done using a high-level evaluation of the overall I&T risk the enterprise faces. In practice, this can be achieved by a review of the components of the business, mission or other components of the risk landscape. The preliminary context provides a perspective on the inherent risk of the enterprise (i.e., an assessment of the I&T risk without taking into account any detailed risk analysis results and, thus, failing to consider existing controls or other risk responses).

Other factors that may be considered include:

- Dependency of the organization on a supply chain, especially one based in another geographic region of the world or reliant upon just-in-time delivery
- The influences of financing, debt, and partners or substantial stakeholders
- Vulnerability to changes in economic or political conditions
- Changes to market trends and patterns
- Emergence of new competition
- Impact of new legislation
- Existence of potential natural disaster
- Constraints caused by legacy systems and antiquated technology
- Strained labor relations and inflexible management
- Geopolitical climate, regulatory or privacy considerations
- Any contractual obligations that would introduce risk to the enterprise if not managed well, such as product liability or liquidated damages clauses

The outcome of the risk scoping activity is used to focus and prioritize more detailed risk management activities. The assessment:

- Allows identification of the potential high-impact risk areas throughout the enterprise
- Provides an overview of major risk factors to which the enterprise is subject, whether or not it has the ability to influence those factors
- Gathers data on any overarching compliance, regulatory, privacy or other obligations (e.g., General Data Protection Regulation [GDPR], Health Insurance Portability and Accountability Act [HIPAA], or country-specific regulations) or contractual obligations that commit the enterprise to specific risk management activities
- Provides first indications of major risk scenarios, which are important inputs to the scenario-building phase of the more detailed risk analysis activities to be performed at a later stage

The enterprise risk scoping activity may need to be repeated on a periodic basis. This can be a simple annual confirmation of earlier results if no major changes have occurred in any of the risk factors, but if major changes (e.g., mergers, new markets) have occurred to the enterprise, the scoping activity should be refreshed. In stable environments, a yearly update or confirmation of the preliminary scope is recommended. It is also recommended to periodically evaluate risk activities to ensure that the most important assets and services are in scope. Some enterprises start with high-value assets that support the most critical business lines, processes or products, or critical services, and then expand the scope as the risk management capability matures.

An enterprise I&T risk scoping exercise involves all major stakeholders of the enterprise. Stakeholders for risk management are identified in the *Risk IT Framework*, chapter 4. The evaluation should be facilitated by an experienced risk management professional to guarantee impartiality and consistency throughout the enterprise.

Data to collect in the scoping activity include:

- The organizational units or subunits, business processes or services, and geographic locations that will be subject to risk management activities
- Impact criteria, e.g., potential consequences of realized risk, and the risk appetite and risk tolerance statements of the enterprise
- Current scoring, severity ratings, or other risk measurement or metrics criteria that are used across the enterprise
- Expected depth and breadth of the risk management activities
- The defined areas that will be subject to reporting and analysis requirements
- The current risk profile of the organizational entity, if one exists
- Risk and control self-assessments that may have been completed
- Audit or compliance priorities

A meeting or discussion, in the form of a table-top exercise or scenario walk-through discussion, using a common concern that many stakeholders may have, is one convenient way to bring together business lines, functional areas and other discrete risk areas. This is especially helpful for enterprises struggling to view both business-line and functional perspectives. For example, if customer data protection is a major interest of the business, a common concern may be the risk of a privacy breach or loss of or destruction of data records. There may also be recurring audit findings related to many functional areas that could be used as an initial case for applying risk management practices across the organization instead of addressing the issue only at one functional, or silo, part of the organization.

A meeting or discussion, in the form of a table-top exercise or scenario walk-through discussion, using a common concern that many stakeholders may have, is one convenient way to bring together business lines, functional areas and other discrete risk areas.

2.1.2 Risk Management Workflow

The risk management workflow consists of the following steps. The steps are not necessarily performed in a linear or sequential manner. Each organization will need to develop a workflow that supports the most efficient and effective means to perform these steps. For ease of reference, **figure 2.1** contains a graphic overview of the steps.

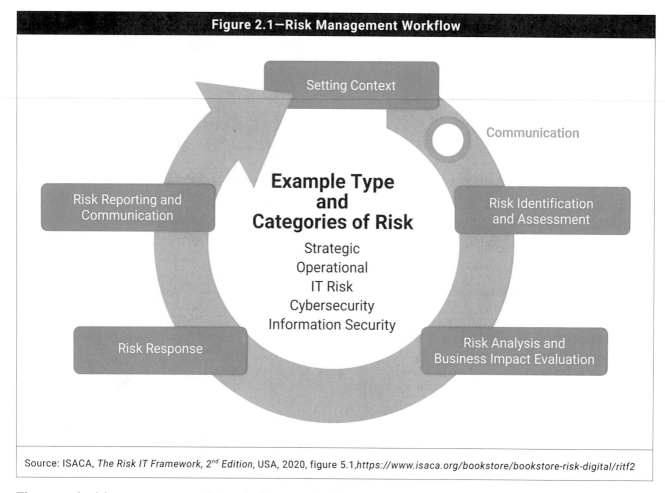

Figure 2.1—Risk Management Workflow

Setting Context

Communication

Risk Reporting and Communication

Risk Identification and Assessment

Example Type and Categories of Risk

Strategic
Operational
IT Risk
Cybersecurity
Information Security

Risk Response

Risk Analysis and Business Impact Evaluation

Source: ISACA, *The Risk IT Framework, 2nd Edition*, USA, 2020, figure 5.1,*https://www.isaca.org/bookstore/bookstore-risk-digital/ritf2*

The example risk management workflow (as illustrated in **figure 2.1**) includes the following steps:

1. As part of setting the context:

    a. Define the scope of the risk assessment—Define the objectives and boundaries of the assessment. Include relevant inputs, ideally derived from or tied to the risk appetite and risk tolerance statements. Perform the assessment in conjunction with the involved business representatives. Business management and staff working in the front line of business or mission activities are often the best sources for areas of concern that may need further inquiry or analysis.

    b. Collect data—Make sure that all possible sources are used to gather relevant data regarding threats, vulnerabilities, conditions, areas of concern, or known risk to business or mission objectives. This includes I&T incident repositories, technology vulnerability reports and change logs, as well as previous risk reports and external data, such as I&T trend analysis and regulatory changes.

2. As part of the risk identification and assessment:

    a. Identify common risk factors—Make sure that interrelated events are grouped by type. Common risk factors can be sourced from a risk taxonomy or a control catalog. These factors can influence frequency and impact of events that may have a significant impact on the business.

    b. Identify areas of concern—Management and staff often have concerns related to tasks and activities that may expose the enterprise or make the enterprise more susceptible to certain types of risk. Sometimes increased exposure to risk may be referred to as the "attack surface" by cyber or information security professionals. This is a good step in which to engage the stakeholders to ensure a thorough identification of all concerns.

3. As part of the risk analysis and business impact:

   a. Analyze I&T risk—Perform the qualitative assessment or quantitative analysis to estimate the frequency and impact of the scenarios, considering the risk factors (which may include current controls). Scenarios and the way to use these are explained in more detail in chapter 4 of this publication. The defined risk tolerance levels (as explained in chapter 3 of this publication) need to be considered as well. These will serve as the basis for determining the risk response. This step needs to be performed with the relevant business representatives and in consultation with the risk oversight functions, if such functions exist.

4. As part of the risk response:

   a. Identify risk response options—This step should be performed in cooperation with the relevant owners of the business processes that depend on the I&T areas that are being assessed.

   b. Understand the current controls or processes that are in place already or need to be designed and how they work to prevent, detect, mitigate, transfer or monitor the risk.

5. As part of risk response and communication:

   a. Review the assessment—The relevant stakeholders on the business side need to be included in the review of the assessment outcome. This will improve the business buy-in to the assessment results and also serve as a way to include their inputs on response actions.

   b. Reporting—Provide management with the results of the assessment and corresponding analysis to support decision-making.

COBIT 2019 takes a similar yet slightly different approach to a risk workflow (**figure 2.2**).[6]

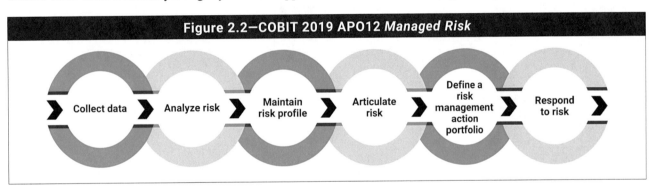

Figure 2.2—COBIT 2019 APO12 *Managed Risk*

Collect data → Analyze risk → Maintain risk profile → Articulate risk → Define a risk management action portfolio → Respond to risk

2.1.3 Example Risk Management Workflow Using a Swimlane (Functional) Diagram

This section depicts a sample risk management workflow in a functional diagram (**figure 2.3**). This functional diagram, also known as a swimlane diagram, references various types of risk management activities and is not the only way to perform the activities. The guidance provided here ties together a number of concepts and activities to assist in making the practices usable in a practical manner. In this context, the following assumptions are used:

- Risk governance is in place and adequately operating in the enterprise. This means there are periodic risk assessments conducted, using an organizationally defined impact scale or set of criteria, and there are defined risk appetite and tolerances that can be used in setting operational thresholds, indicators or triggers used to initiate risk management activities.

- Risk assessment and analysis criteria are used to update the risk register, risk maps and the risk profile, if there is one.

Risk is proactively identified, using a variety of techniques and methods, and there is a way to intake or accept the reporting of an area of concern so it can be investigated further.

---

[6] For activities that support the COBIT 2019 risk workflow, see ISACA, *COBIT® 2019 Framework: Governance and Management Objectives*, APO12 *Managed risk*.

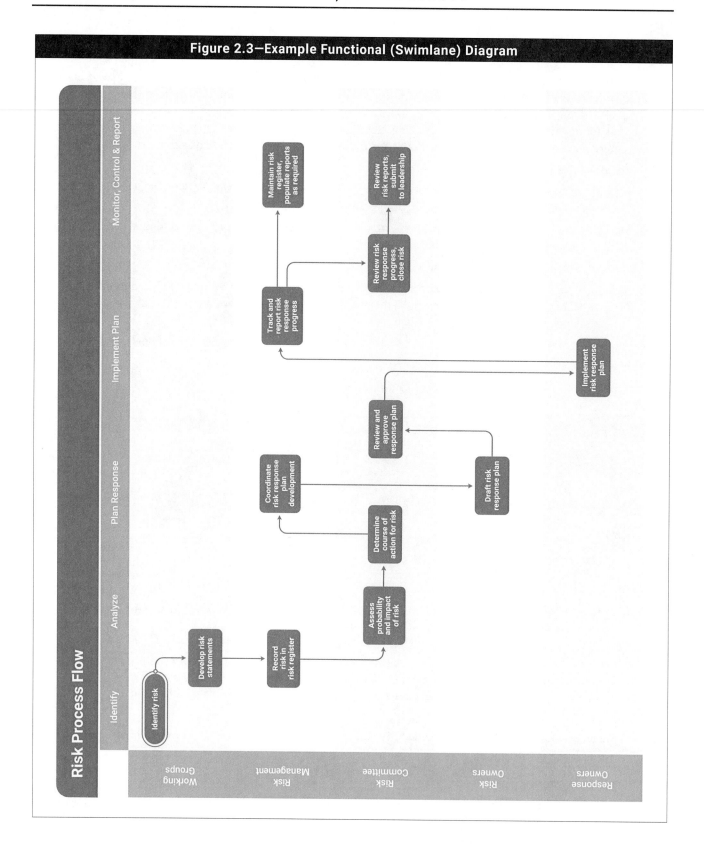

Figure 2.3—Example Functional (Swimlane) Diagram

Chapter 3
Essentials of Risk Governance

3.1 Governance

The term "governance" has moved to the forefront of business thinking in response to examples demonstrating the importance of effective oversight on one end of the spectrum and the global business mishaps derived from poor monitoring on the other. Corporate governance is the system by which organizations are evaluated, directed and controlled. By implication, the corporate governance of I&T is the system by which the current and future use of I&T is evaluated, directed and controlled. The objective of any governance system is to enable organizations to create value for their stakeholders or to promote value creation. Value creation, in turn, comprises benefits realization, risk optimization and resource optimization. Risk optimization is an essential part of any governance system and cannot be seen in isolation from benefits realization or resource optimization. For those readers who are familiar with COBIT, the content in this section is compatible with COBIT 2019 governance objective EDM03 *Ensured risk optimization*.[7]

Governance ensures that stakeholder needs, conditions and options are evaluated to determine balanced, agreed-on enterprise objectives to be achieved; direction is set through prioritization and decision-making; and performance and compliance and progress against agreed-on direction and objectives are monitored. In most enterprises, governance is the responsibility of the board of directors under the leadership of the chairperson, as shown in **figure 3.1**.

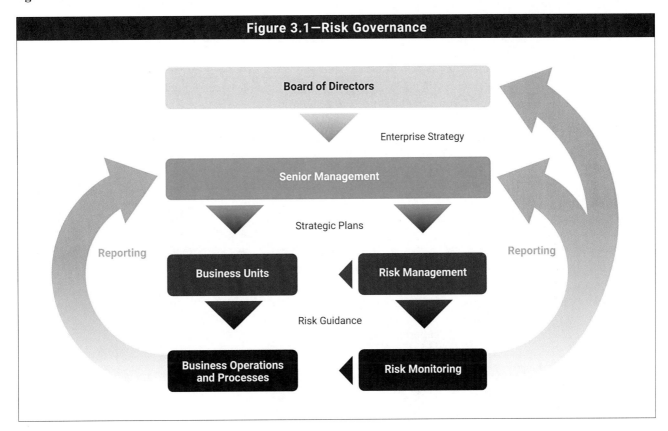

Figure 3.1—Risk Governance

---

[7] See ISACA, *COBIT® 2019 Framework: Governance and Management Objectives*, EDM03 *Ensured risk optimization*.

Good governance means that risk optimization is part of the arrangements that are put in place, and risk information is included in the decision-making process. At the same time, the risk function needs to be governed, that is, provided with direction and monitored.

The topics discussed in this chapter include:

1. Risk appetite and risk tolerance
2. Risk culture
3. Risk policy
4. Key risk indicators (KRIs)
5. Risk profile
6. Risk maps and risk aggregation for board and executive decision-making

The topic of risk capacity is covered in the *Risk IT Framework* (published separately) and not repeated in this document.

3.1.1 Risk Appetite and Risk Tolerance

In formulating strategies and operating plans, an enterprise is exposed to some level of risk to achieve its objectives. The amount of risk is generally expressed as risk appetite and risk tolerance. In COBIT, this is referred to as optimization, which means keeping risk within tolerance to the risk appetite, which should be the goal. Risk appetite and tolerance are concepts that are frequently used, but the potential for misunderstanding is high. Some people use the concepts interchangeably, others see a clear difference. Risk appetite can and will be different in each enterprise—there is no absolute norm or standard of what constitutes acceptable and unacceptable risk.

Effective risk management begins at the highest levels of the organization, with well-formed and clearly articulated risk appetite statements. The statements, when clearly understood, communicated and practiced, serve as the guide to the behaviors, decisions, limits and policies that provide the boundaries under which risk management practices operate within an enterprise.

As a reminder, risk appetite is the amount of risk an entity—an enterprise, organization, public or private company—is willing to take to achieve its strategic objectives. For example, a risk appetite statement for a healthcare provider might be, "We place patient safety as our top priority. We also recognize the need to balance the level of immediate response to all patient needs with the cost of providing such service." This demonstrates a low appetite for risk that might impact patient safety balanced with a higher appetite related to response to patient care and customer service.

Risk tolerance is the amount of variation in the parameters used to measure risk appetite. In other words, risk tolerance is the response to the specific risk appetite. For the healthcare provider noted in the preceding paragraph, example risk appetite and corresponding tolerance statements might be, "We plan our staffing levels to enable treatment of all patients within five minutes of their appointment time and emergency walk-in patients within 15 minutes. However, management accepts that in rare situations (5 percent of the time), patients in need of non-life-threatening attention may not receive that attention for up to four hours."

Risk tolerance is the amount of variation in the parameters used to measure risk appetite.

Here are some tips for evaluating risk appetite and considerations for developing statements that can be tested and improved over time:

- Are the management and governance entities of the organization aligned on the business outcomes that are unacceptable to the enterprise? What is the process to periodically evaluate the enterprise's risk appetite statements if there are significant changes in its business, mission or other conditions?

- Are the unacceptable outcomes clear and communicated to everyone who needs to know? Is everyone clear on the types of risk the enterprise wants to take vs. those it wants to avoid?

- If someone becomes aware of a potential risk is there a way to raise a concern or ask for an inquiry before a negative event occurs? How would a determination be made of the effectiveness of the organization's process for identifying, assessing and reporting risk in relation to the stated risk appetite?

- Do the people on the front line of the enterprise know what the boundaries, parameters, control limits or other constraints on risk-taking decisions are for their role? Examples of control limits in a bank might be related to the financial limit granted to various staff members to cash a check—a junior staff member's upper financial limit being lower than that of a senior staff member, whose higher limit is based on his/her greater experience and training in the operational environment. In many businesses, there are upper and lower limits on decision-making that are integrated into the workflow to assist in making sure the optimal risk appetite is maintained.

- Are there published financial loss limits; regulatory compliance; business interruption; operational performance; life, health, or safety impacts that are clearly defined and communicated? Do these published limits exist for information security, cybersecurity, and technology events or incidents?

Raising awareness about risk in an organization is about acknowledging that uncertainty, or risk, is an integral part of the business. This does not imply that all risk is to be avoided or eliminated, but rather that it is effectively communicated and well understood by all. I&T risk is identifiable, and the enterprise that uses a common language for expressing and describing risk will more easily be able to detect, recognize, and use appropriate resources to manage it.

Raising awareness about risk in an organization is about acknowledging that uncertainty, or risk, is an integral part of the business.

Defining and developing an enterprise set of risk appetite and risk tolerance statements are only part of the success criteria for risk management. A commonly used language is also needed so that all stakeholders are able to receive, understand and act in accordance with the policies and activities needed to ensure that no risk materializes that would prevent the mission or business objectives from being attained. Risk reporting and communication are key parts in this process; it is critical for decision-makers and stakeholders (including boards) to receive timely and accurate information on risk that can be acted upon.

Talking about risk may cause some uncomfortable conversations since risk involves forward-looking uncertainty and may not actually ever be realized. However, it is imperative that good risk communication is practiced before it is realized as an issue, incident or major crisis.

There are multiple options for expressing I&T risk in business terms, and there is no right or wrong option. It is critical to choose the option that fits best with the enterprise and complement this scheme with a range of scales to quantify the risk during risk analysis.

This risk appetite statements and expressions of tolerance in **figure 3.2** are examples. Every enterprise has to define its own risk appetite levels and review them on a regular basis. This definition should be in line with the overall risk culture that the enterprise wants to express (i.e., ranging from very risk averse to risk taking/ opportunity seeking). Risk appetite and risk tolerance should be applied to all I&T decision-making.

Figure 3.2—Sample Enterprise Risk Appetite and Tolerance Statements		
Type of Enterprise	Risk Appetite Example Statement	Risk Tolerance Example Statement
University or higher learning institute	The university system is willing to assume credit interest rates of X percent for borrowing (a certain financial amount or percentage of assets) to fund new initiatives.	The university credit rating may not drop more than one grade from its current level.
University or higher learning institute	The university system accepts an investment of US $X per headcount in recruiting and training for new employees.	On a universitywide basis, employee turnover is to be less than X percent in any given 90-day period.
Energy or utility provider	The reputation and financial condition of the enterprise could be negatively impacted due to the enterprise's obligations to comply with federal and state regulations, laws and other legal requirements that govern operations, assessments, storage, closure, remediation, disposal and monitoring compliance.	On an enterprisewide basis, noncompliance penalties are to be less than X percent in a 12-month period.
Energy or utility provider	The enterprise strives to maintain 99.999 percent availability of power to its customers and to restore power to customers within X hours of being notified of a service outage.	Service outages to more than X percent of customers for less than (Y-time period) are acceptable.
Financial institution	The bank has a low appetite for IT system-related incidents that are generated by poor change management practices.	X percent of the bank's technology assets must have approved configuration settings that have been established and implemented and are maintained. Another change management example tolerance statement might establish an approved percentage of successful changes to critical business assets during a specified time period.
Financial institution	The bank is committed to ensuring that its customer information is correct, appropriately classified, properly protected and managed in accordance with legislative and business requirements.	The bank will maintain (X) frequency and timeliness of information asset backups and successful testing of backups. Another customer information tolerance statement might establish an allowed number of policy violations related to confidentiality and privacy of customer information.

The example risk map in **figure 3.3** is a means to depict risk on a two-dimensional graph, using the dimensions of frequency and impact. Risk appetite can be visualized using the same risk maps—different bands of risk significance can be defined, indicated by colored squares on the risk map.

Risk acceptance is binary: Something is, or is not, acceptable. In practice, this is about how quickly an organization responds to a risk-related condition. It may be that no response is necessary because the condition is well within defined tolerances (i.e., it is acceptable or represents an opportunity to take additional risk), a response is required because a condition is either outside of tolerance or headed in that direction (i.e., it is unacceptable but not urgent), or a condition is of a nature that requires immediate response because it is far outside of tolerance.

In the example in **figure 3.3**, four levels of significance are defined:

Red indicates unacceptable risk that requires immediate action.

Yellow indicates unacceptable risk (also above risk appetite).

Green indicates an acceptable level of risk with no special action required.

Blue indicates very low risk. This may be an area where efficiencies or cost-saving opportunities might be found by decreasing the degree of control, or assuming more risk might be indicated.

Figure 3.3—Example Risk Map

	Frequency		
Impact	Unacceptable	Unacceptable—Immediate action required	Unacceptable—Immediate action required
	Acceptable	Unacceptable	Unacceptable—Immediate action required
	Opportunity	Acceptable	Unacceptable

Typically, multiple areas of risk exist. As a result, it is helpful for senior management to have view of where each area falls in relation to the overall risk appetite of the enterprise. **Figure 3.4** is an example of how risk can be plotted within bands based on the impact and frequency of each risk.

Figure 3.4—Risk Map Indicating Risk Appetite Bands

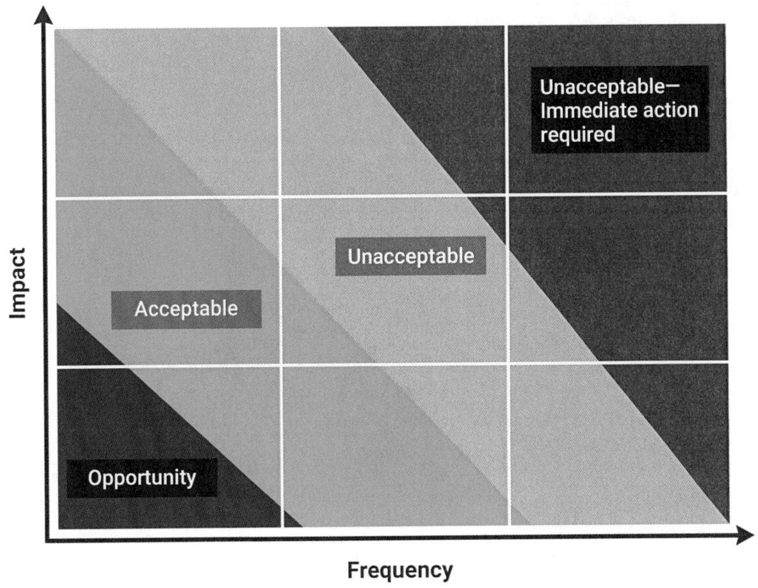

This risk appetite scheme is an illustrative example (**figure 3.5**). The letters denote various risk that may exist within an enterprise. Risk appetite can be explained and displayed via a Risk Map.

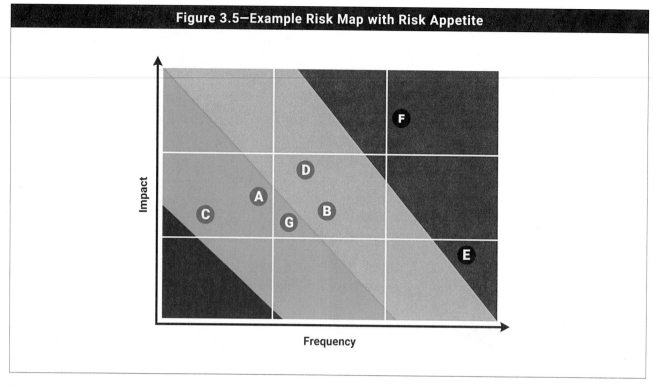

Figure 3.5—Example Risk Map with Risk Appetite

Current practice and qualitative risk assessment methods may not be mature enough to provide a level of precision as illustrated in **figure 3.5**.

Risk appetite is defined by senior management at the enterprise level.[8] There are several benefits associated with defining risk appetite at the enterprise level. This approach:

- Supports and provides evidence of the risk-based decision-making process
- Tracks and presents the decisions for each type of risk response
- Supports the understanding of how each component of the enterprise contributes to the overall risk profile
- Shows how different resource allocation strategies can have an impact on the risk response options and desired outcomes for risk response options
- Enables risk response actions to be tracked back and justified
- Supports consistency in risk management decisions
- Supports the prioritization of risk response actions
- Identifies specific areas where risk should receive a response

An enterprise's risk appetite and tolerance change over time. New technology, new organizational structures, new market conditions, new business strategy and many other factors require the enterprise to reassess its risk portfolio at regular intervals. These developments also require the enterprise to reconfirm its risk appetite at regular intervals and may trigger risk policy reviews.

Risk appetite and risk tolerance go hand in hand. Risk tolerance is defined at the enterprise level and is reflected in policies set by the executives. At lower (tactical) levels of the enterprise, exceptions can be tolerated (or different thresholds defined) as long as, at the enterprise level, the overall exposure does not exceed the desired risk appetite.

---

[8] *Ibid.*

Any business initiative includes a risk component, so management should have the discretion to pursue new opportunities up to the level of risk appetite. There are situations in which policies are based on specific legal, regulatory or industry requirements, rendering it appropriate to have no risk tolerance for failure to meet the requirement. An example of this may be related to specific privacy regulations in certain countries where noncompliance may result in extremely large financial fines or penalties.

3.1.2 Risk Culture

Risk management is about helping an enterprise maximize the benefits it is able to generate, while avoiding losses that can negatively impact its ability to achieve its mission or its ability to continue operations. A risk-aware culture characteristically offers a setting in which components of risk are discussed openly, and acceptable levels of risk are understood and maintained. A risk-aware culture begins at the top, with board and business executives who set direction, communicate risk-aware decision-making and reward effective risk management behaviors. Risk awareness also implies that all management and staff of any level within an enterprise are aware of how and why to respond to adverse I&T events.

Risk culture is a concept that is not easy to describe. It consists of a series of behaviors, as shown in **figure 3.6**.

Figure 3.6—Relevant Behavior for Risk Governance and Management	
Behavior	**Key Objective/Suitable Criteria/Outcome**
General Behavior	
A risk- and compliance-aware culture exists throughout the enterprise, including the proactive identification and escalation of risk.	The culture must define a risk management approach and risk appetite. Zero tolerance of noncompliance with legal and regulatory requirements must be established.
Defined policies are in place that have been communicated and that drive behavior.	All personnel understand and implement the requirements of the enterprise as defined in policies.
The enterprise shows positive behavior toward raising issues or negative outcomes.	Whistleblowers are seen as having a positive contribution to the enterprise. A blame culture is avoided. Personnel understand the need for risk awareness and reporting possible weaknesses.
The enterprise recognizes the value of risk.	Personnel understand the importance to the enterprise of maintaining risk awareness and the value that managing risk adds to their role.
The enterprise fosters a transparent and participative culture as an important focus point.	Communication is open and overt so facts are not omitted, misrepresented or understated. The negative impact of hidden agendas is avoided.
Stakeholders show mutual respect.	Stakeholders and risk assessors are encouraged to collaborate. People are respected as professionals and treated as experts in their roles.
The business accepts ownership of risk.	Risk practices are incorporated throughout the enterprise. Accountabilities are cleared and accepted. IT-related business risk is owned by the business and not viewed as solely the responsibility of the IT department for the risk function.
The enterprise allows risk acceptance as a valid option.	Management understands the likelihood and consequence of the impact of risk acceptance. The impact is determined to be within the enterprise's risk appetite.
Risk Professional Behavior	
Effort is made to understand what risk is for each stakeholder and how it impacts the stakeholder's objectives.	Risk professionals understand the commercial reality of the impacts of risk. This may include competitive, operational, regulatory and compliance requirements. Although there may be risk common to a certain industry, each enterprise is unique in terms of how these risk items impact specific enterprise objectives.

Figure 3.6—Relevant Behavior for Risk Governance and Management *(cont.)*	
Behavior	**Key Objective/Suitable Criteria/Outcome**
Risk professionals create awareness and understanding of the risk policy.	Alignment among risk capacity, risk appetite and enterprise policy can lead to effective risk strategies.
Collaboration and two-way communication during risk assessment are supported.	Risk assessment is fundamentally accurate and complete and it addresses stakeholder needs.
The enterprise's risk appetite is clear and communicated in a timely fashion with relevant stakeholders.	Stakeholders manage risk more effectively and there is appropriate alignment with organizational strategy and efforts.
Policies reflect risk appetite and risk tolerance.	Employees and management operate within risk tolerance. Business lines apply formal risk appetite and tolerance to daily practices. There is a clear process for proposing and making changes to risk appetite levels, with senior management consideration and approval.
Enterprise culture supports effective risk practice.	Stakeholders understand risk from common portfolio views (product, process) and apply risk-based decision-making to daily practice.
KRIs are used as an early warning.	KRIs are associated with valid metrics and can be used as an indicator of process or control failure. KRI metrics are available and accessible for regular reporting and relate to objectives.
Risk indicators or events that fall outside of tolerance and appetite are acted on.	Risk indicators are linked to the management risk response and remediation activities.
Management Behavior	
Senior management sets direction and demonstrates visible and genuine support for risk practices.	Quality risk management practices are maintained through genuine support from senior management.
Management engages with all relevant stakeholders to agree on actions and follow up on action plans.	The correct stakeholders are appropriately involved in ensuring timely resolution of issues and the achievement of business plans.
Genuine commitment is obtained and resources are assigned for execution of actions.	Personnel are empowered in executing actions required by risk management decisions.
Management aligns policies and actions to risk appetite.	Management makes appropriate risk decisions in complying with policies. Risk-adjusted revenue is in line with management expectations.
Management proactively monitors risk and action plan progress.	Remediation plans are completed within expected business time frames and have a positive impact on enterprise objectives.
Risk trends are reported to management.	The timely reporting of risk trends proactively manages risk and avoids lost opportunities.
Effective risk management is rewarded.	Good risk practice is acknowledged. Employees' performance goals and reward structures are set to stimulate effective risk management practices and appropriate execution of mitigation actions.

Risk culture includes:

- Behavior toward taking risk—What are the norms and attitudes toward taking risk, identification of risk and analysis of risk?

- Behavior toward following policy—Is policy something that exists but is not followed? Do policies drive behavior? Are policies easy to read, understand and follow?

- Behavior toward negative outcomes—How does the enterprise deal with negative outcomes, policy exceptions, loss events, cyberincidents, missed opportunities, audit findings and incident after-action investigations? Will it learn from them and try to adjust, or will blame be assigned without treating the root cause?

Some symptoms of an inadequate or problematic risk culture include:

- Misalignment among risk appetite, stated tolerances and translation into policies, guidance when policies are not aligned with management direction, and organizational norms regarding compliance with policy

- Large numbers of policy exceptions, which suggest that either policies and standards do not, in fact, represent the organization's risk appetite/tolerance or that the organization does not properly vet exception requests.

- Existence of a blame culture. This type of culture should be avoided as it is the most effective inhibitor of relevant and efficient communication. In a blame culture, business units tend to point the finger at each other when objectives are not met. In doing so, they fail to realize how the business unit's involvement up front affects project success. In extreme cases, the business unit may assign blame for a failure to meet the expectations that the unit never clearly communicated. The blame game only detracts from effective communication across units, further fueling delays. Executive leadership must identify and quickly control a blame culture if collaboration is to be fostered throughout the enterprise.

3.1.3 Risk Policy

Good practice in risk management requires that policies be part of an overall governance and management framework, providing a (hierarchical) structure into which all policies should nest and provide support to the underlying principles.

As part of including risk management norms or conditions into the enterprise policy framework, the following items should also be described in risk policies:

- Scope and authority, tied to risk appetite or tolerance statements
- Roles and responsibilities of the stakeholders
- The consequences of failing to comply with the policy
- The means for handling exceptions
- The manner in which compliance with the policy will be checked and measured

Policies should be aligned with the enterprise's risk appetite. Policies are a key component of an enterprise's system of internal control, whose purpose it is to ensure that an enterprise meets its stated objectives. As part of risk governance activities, the enterprise's risk appetite is defined and this risk appetite should be reflected in the policies. This is not meant to suggest that risk appetite or risk tolerance statements be embedded into policy documents, rather that the policies need to be aligned with the risk-taking culture of the enterprise. Policies need to be revalidated and updated at regular intervals to ensure relevance to business requirements and practices.

As part of risk governance activities, the enterprise's risk appetite is defined and this risk appetite should be reflected in the policies.

Policies provide detailed guidance on how to put principles into practice and how they will influence decision-making. Example risk policy types are listed in **figure 3.7**. Not all relevant policies are written and owned by the IT, information security, information privacy or risk function.

Figure 3.7—Example Risk Policy Types	
Policies	**Description**
Core IT risk policy	Defines, at strategic, tactical and operational levels, how the risk of an enterprise needs to be governed and managed pursuant to its business objectives. This policy translates enterprise governance into risk governance principles and policies and elaborates risk management activities.
Information security policy	Establishes rules for protecting corporate information and the associated systems and infrastructure. The business requirements regarding security and storage are more dynamic than IT risk management, so, for effectiveness, their governance needs to be handled separately from the governance of IT risk. However, for operational efficiency, it is necessary to keep the information security policy in sync with the IT risk policy.
Crisis management policy	As with IT security, network management and data security, IT crisis management is one of the operational level policies that needs to be considered for complete IT risk management. It sets the guidelines on how to act in situations of crisis and details the sequence in which to deal with each of the identified (key) areas of risk.
Third-party IT service delivery management policy	Establishes guidelines for managing the risk related to third-party services. It sets out a framework of expectations in behavior and security precautions taken by third-party service providers to manage the risk related to the service provision.
Business continuity policy	Contains management's commitment and view on: • Business impact analysis (BIA) • Business contingency plans with trusted recovery • Recovery requirements for critical systems • Defined thresholds and triggers for contingencies • How to handle escalation of incidents • Disaster recovery plan (DRP) • Training and testing
Program/project management policy	Deals with managing risk linked to projects and programs. It details management's position and expectation regarding program and project management. Moreover, it handles accountability, goals and objectives regarding performance, budget, risk analysis, reporting and mitigating adverse events during the execution of programs and projects.
Human resources (HR) policies	Detail what employees can expect from the enterprise and what the enterprise expects from employees. They provide detailed acceptable and unacceptable behavior by employees and, in doing so, manage the risk that is linked to human behavior.
Fraud risk policy	Is concerned with protecting the enterprise brand, reputation and assets from loss and/or damage resulting from incidents of fraud and/or misconduct. The policy provides guidance to all employees on reporting any suspicious activities and ways to handle sensitive information and evidence. It helps to raise an antifraud culture and awareness of risk.
Compliance policy	Explains the assessment process regarding compliance with regulatory, contractual and internal requirements. It lists roles and responsibilities for the different activities in the process and provides guidance on metrics to be used to measure compliance.
Ethics policy	Defines the essentials of how people within an enterprise will interact with one another and with any customers or clients they serve
Quality management policy	Details management's vision on the quality objectives of the enterprise, the acceptable level of quality and the duties of specific departments to ensure quality
Service management policy	Provides direction and guidance to ensure the effective management and implementation of all IT services to meet business and customer requirements, within a framework of performance measurement. It also deals with management of risk related to IT services. Detailed guidance on service management and optimization of risk related to services is included in the ITIL framework.

Figure 3.7—Example Risk Policy Types *(cont.)*	
Policies	**Description**
Change management policy	Communicates management's intent that changes to the enterprise IT be managed and implemented in a way that minimizes risk and impact to the stakeholders. The policy contains information on the assets in scope and the established standard change management process.
Delegation of authority policy	Details: • The authority that the board retains strictly for itself • The general principles of delegation of authority • A schedule of the delegation of authority (including clear boundaries) • A clear definition of the organizational structures to which the board delegates its authority
Whistleblower policy	Should: • Encourage employees to raise concerns and questions • Provide avenues for employees to raise concerns in full confidence • Ensure employees that they will receive a response to raised concerns and will be able to escalate a concern if they are not satisfied with the response • Reassure the employees that they are protected when they raise issues and should not fear reprisals
Internal control policy	The purpose is to: • Communicate management's internal control objectives • Establish standards for the design and operation of the enterprise system of internal controls to reduce the exposure to all risk faced by the enterprise
Intellectual property (IP) policy	Ensures that all risk related to the use, ownership, sale and distribution of the outputs of IT-related creative endeavors by employees of an enterprise, such as software development, is detailed in an appropriate way, from the start of an endeavor
Data privacy policy	Describes the ways that a party gathers, uses, discloses and manages personal data. Personal information can be anything that can be used to identify an individual, including but not limited to name, address, date of birth, marital status, contact information, ID issue and expiry date, financial records, credit information, medical history, travel destination, and intentions to acquire goods and services. The policy defines how an enterprise collects, stores and releases the personal information that it collects. It informs the client of the specific information that is collected and whether it is kept confidential, shared with partners, or sold to firms or enterprises. Furthermore, the policy ensures compliance with relevant legislation related to data protection.

3.1.4 Key Risk Indicators (KRIs)

Any measurement that can be used to describe and track a risk is an indicator of that risk. Risk indicators are specific to each enterprise. Development and selection of KRIs depend on a number of parameters in the internal and external environment, such as the size and complexity of the enterprise, whether it is operating in a highly regulated market and its strategic objectives.

Identifying risk indicators should take into account the following aspects:

• Consideration of the different stakeholders in the enterprise. Risk indicators can and should be identified for stakeholders depending on their unique information needs. Involving the right stakeholders in the selection of risk indicators will also ensure greater buy-in and ownership.

- Making a balanced selection of risk indicators, covering lag indicators (indicating risk after events have occurred), lead indicators (indicating what capabilities are in place to prevent events from occurring) and trends (analyzing indicators over time or correlating indicators to gain insights). The selected indicators must drill down to the root cause of the events (indicative of root cause and not just symptoms).

- Indicators that tie directly to the stated risk appetite and risk tolerance statements will be most meaningful to track desired outcomes of the risk management program.

KRIs are the metrics or pieces of data that serve as an early warning signal that something is not operating as expected or there is increased risk exposure in one or more areas of the enterprise. This type of indicator is similar to a smoke detector that sounds an alert at the first hint of smoke. An indicator that meets the early warning requirement is known as a leading indicator. A leading indicator may not be completely reliable or there may be alerts that turn out to be false positive indicators.

No discussion of risk indicators would be complete without introducing key performance indicators (KPIs). KPIs are designed to provide a high-level overview of past performance and are almost always derived from historical data. KPIs are known as lagging indicators and provide information on whether or not targets were met for a compliance requirement or expenses were controlled on a project at the completion of the project.

Using both KRIs and KPIs when getting started with risk management measurements may be necessary until there are enough data to improve the process. Ideally, KRIs are leading indicators but that may not always be possible in the early stages of risk measurement. Examples of KRIs are shown in **figure 3.8**.

Figure 3.8—Example Key Risk Indicators
Percent of critical business areas that have completed a risk assessment in the past 12 months
Percent of compliance or controls testing regime for critical assets completed in past 12 months
Percent of unmitigated risk in the risk register for which no response plan exists
Percent of business processes in which customer personal data are used
Percent of incidents in which customer personal data are lost/stolen
Percent of lost or delayed product deliveries due to IT failure
Percent of suppliers providing critical IT systems who have implemented an approved security control framework
Percent of DRPs successfully tested in the past 12 months (lagging)
Percent of DRPs scheduled to be tested in the next 12 months (leading)

An enterprise may develop an extensive set of metrics to serve as risk indicators; however, it is neither possible nor feasible to maintain that full set of metrics as KRIs. A KRI is differentiated as being highly relevant and possessing a high probability of predicting or indicating important risk. Criteria to select KRIs include:

- **Impact**—Indicators for risk with high business impact or those tied directly to risk tolerance statements are more likely to be KRIs.

- **Effort** (to implement, measure and report)—For different indicators that are equivalent in sensitivity, the one for which data collection is easier may be a good starting point.

- **Reliability**—The indicator must possess a high correlation with the risk and be a good predictor or outcome measure.

- **Sensitivity**—The indicator must be representative of the risk factors and capable of accurately indicating variances in the risk factors.

Risk that is to be entered into the risk management workflow is then monitored, reported upon or closed when the risk is deemed to be in an acceptable range for the enterprise. In this stage of risk management, two common methods used for developing risk indicators, some of which might result in KRIs, are root cause analysis (RCA)[9] and Goal-Question-Indicator-Metric (GQIM).[10] Both of these methods help with the development of indicators and improve the feedback loop that is needed to mature the risk management process over time.

3.1.5 Maps and Risk Aggregation for Board and Executive Decision Making

A very common and intuitive technique to present risk is the risk map, also known as a heatmap, where risk is plotted on a two-dimensional diagram, with frequency and impact being the two most common dimensions (X and Y axis). A sample risk map is shown in **figure 3.9**. The map may be helpful to provide a quick prioritization triage exercise but may not give management enough information to take a decision on appropriate action.

Heatmaps, although in wide use in many enterprises, are often the subject of much criticism and dismissed by many quantitative risk analysts as having little value, with some analysts referring to them as practically useless. As further described in the qualitative and quantitative risk analysis sections later in this publication, the risk map should be an expression of risk that has been evaluated using well-defined and unambiguous impact criteria. Initial maps or an aggregation of risk types may be useful for an analyst to develop trends or common profiles for which risk response activities could be made more efficient, but they may not be the right technique for management decision-making. Most of the time, if an organization is using only qualitative risk assessment methods it is very difficult, if not impossible, to aggregate risk in a meaningful way.

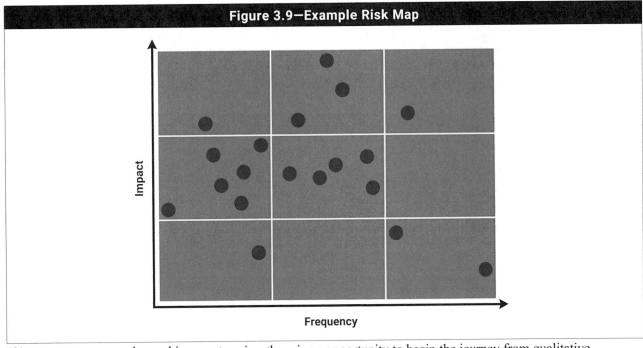

Figure 3.9—Example Risk Map

If heatmaps are currently used in an enterprise, there is an opportunity to begin the journey from qualitative "measurement" of risk to quantitative measurement of risk. For example, it is possible to start the journey with some simple steps, such as to moving from qualitative descriptions of risk (high, medium or low) by associating these words with specific numbers or, more accurately, specific ranges of numbers.

Once the enterprise risk analysts and stakeholders are comfortable using and communicating with the basic risk map, the concept can be extended to more advanced maps that provide additional quantitative criteria for aggregation of risk and prioritized decision-making (**figure 3.10**). This is a rapidly changing and adaptive area in the field of risk

9   Durmesevic-Mutapcic, A.; "How Root Cause Analysis Fits Into Various Audit Types," *ISACA® Journal*, 1 March 2019, *https://www.isaca.org/resources/isaca-journal/issues/2019/volume-2/how-root-cause-analysis-fits-into-various-audit-types*
10 Stewart, K.; Allen, J.; Valdez, M.; Young, L; "Measuring What Matters Workshop Report," Software Engineering Institute, CERT Division, January 2015, *https://resources.sei.cmu.edu/asset_files/TechnicalNote/2015_004_001_433525.pdf*

management that will be addressed by several new tools, techniques and methods in the future. An article published in the *ISACA Journal*, titled "Evolving from Qualitative to Quantitative Risk Assessment," provides additional information and relevant examples.[11]

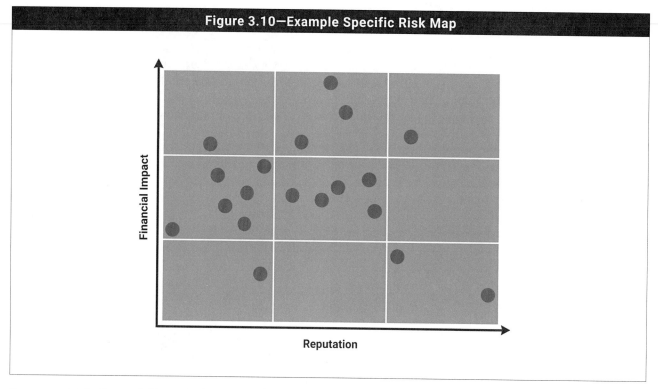

Figure 3.10—Example Specific Risk Map

For example, in **figure 3.10**, analysis was performed on the risk factors related to the direct financial cost (X axis) that may be incurred by a data breach occurring, with the Y axis replaced with a frequency count of the number of times the organization was mentioned in the news or other media.

If the board and management have a requirement to quantify risk in financial terms, the most common techniques are to shift the focus to the calculation of the probable maximum loss (PML)[12] or the maximum foreseeable loss (MFL).[13] Using PML or MFL allows the risk to be analyzed with regard to the impact on the organization should the risk materialize. These techniques are used for ensuring the enterprise has the appropriate financial protections in place should a risk materialize.

The financial impact of risk is often aggregated for executive or board reporting purposes into ranges of monetary loss that could be expected if certain risk types were realized. In many organizations there is a set of impact criteria and risk tolerances expressed in financial terms. Often this visualization of risk is also displayed on a risk map, using various sizes of circles to represent the financial impact: a large circle representing a higher financial impact than a smaller circle, as shown in **figure 3.11**.

The financial impact of risk is often aggregated for executive or board reporting purposes into ranges of monetary loss that could be expected if certain risk types were realized.

---

[11] Heynderickx, B.; "Evolving From Qualitative to Quantitative Risk Assessment," *ISACA Journal*, 1 July 2019, *https://www.isaca.org/resources/isaca-journal/issues/2019/volume-4/evolving-from-qualitative-to-quantitative-risk-assessment*
[12] Investopedia, "Probable Maximum Loss," *https://www.investopedia.com/terms/p/probable-maximum-loss-pml.asp*
[13] Investopedia, "Maximum Foreseeable Loss," *https://www.investopedia.com/terms/m/maximum-foreseeable-loss.asp*

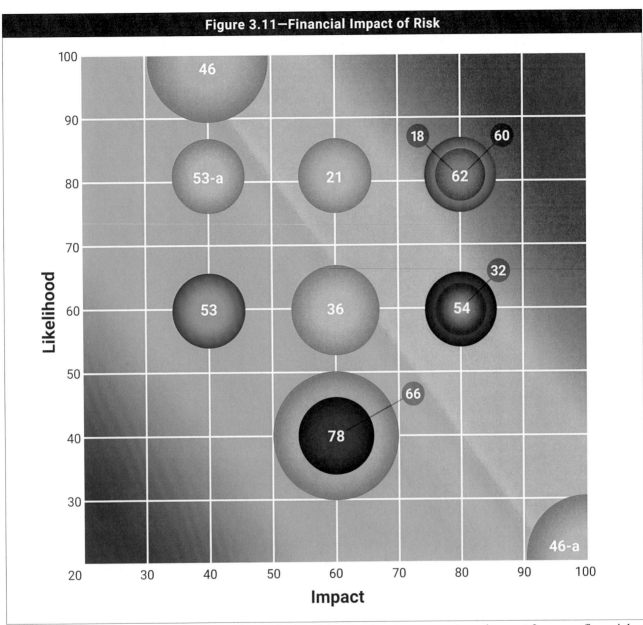

Figure 3.11—Financial Impact of Risk

In the financial sector, risk aggregation and quantitative risk reporting are a current requirement for many financial institutions subject to the Basel Committee on Banking Supervision (Basel Committee) supervisory process.[14] This requirement is driving a discussion on the best methods for risk aggregation and risk quantification so that I&T risk can be reported and managed commensurately with other risk that requires informed decision-making.

Aggregation allows risk that spans across an entire enterprise to become visible and informs management's decision-making on risk responses. Risk, when aggregated at an enterprise level, may be so far out of the range of organizational tolerance that immediate action or widespread responses are required. Risk aggregation is the main driver for developing and refining risk impact criteria that reflect the organizationally specific and common scales of measurement needed to truly manage I&T risk in a way that is commensurate with other enterprise risk. Maturing the risk management functions and processes to the level of capability that allows more certainty in decision-making creates greater value for the organization than just managing a single risk. Informed decision-making is truly the value of a mature quantitative risk management process.

---

[14] BIS, "Launch of the consolidated Basel Framework," 16 December 2019, *https://www.bis.org/bcbs/publ/d491.htm*

Page intentionally left blank

Chapter 4
Essentials of Risk Assessment

4.1 Essential Components

This chapter discusses the essential components of risk assessment. Enterprise management is responsible for the plan, build, implementation and monitoring activities in alignment with the direction set by the governance body to achieve enterprise objectives. In most enterprises, management is the responsibility of executive management, under the leadership of the chief executive officer (CEO).

As explained in chapter 3, governance is the system by which organizations are evaluated, directed and controlled. There is a clear distinction between governance and management. Management focuses on planning, building, executing and monitoring activities in alignment with the direction set by the governance body to create value by achieving objectives. A well-managed organization subject to poor governance will create and execute clear, effective plans to attain objectives that do not create value. Similarly, risk management attempts to foresee the challenges to achieving objectives and lower the probabilities of negative outcomes occurring (or their impacts if they do occur), but the effectiveness of risk management depends in large part on decisions made by those responsible for risk governance. For those practitioners who are more familiar with COBIT 2019 the activities in this chapter are related to APO12 *Managed risk*.[15]

The topics discussed here include:

1. Risk criteria: expressing impact in business terms
2. Risk identification
3. Risk assessment and analysis
4. Qualitative and quantitative approaches
5. Risk maps (heatmaps)
6. I&T risk scenarios
7. Risk register

4.1.1 Risk Criteria: Expressing Impact in Business Terms

Risk management is an enterprise activity that benefits from a standardized, structured approach that can be applied to the entire enterprise without substantial modification or customization. It is possible to identify risk on a system-by-system or project-by-project basis, but the result of such an approach creates the risk of false assurance by having neither consistency nor interoperability among the risk methods, techniques or processes that are implemented. Without a structured approach, risk may be measured differently in different parts of the organization, creating confusion and resulting in the management of individual risk rather than the management of risk across the enterprise.

Risk management is an enterprise activity that benefits from a standardized, structured approach that can be applied to the entire enterprise without substantial modification or customization.

One analysis technique that is useful in this phase is the development of organization-specific impact criteria.

Impact criteria give risk meaning and help the enterprise to set organization-specific risk appetites or risk tolerances. Criteria may be qualitative in nature or quantitative. One reason for developing impact criteria with defined

---

[15] See ISACA, *COBIT® 2019 Framework: Governance and Management Objectives*, APO12 *Managed risk*.

quantitative ranges (thresholds) is that the criteria can act as inputs to or double-checks of the risk appetite and risk tolerance statements.

Impact criteria apply to the enterprise, not a specific I&T asset, and reflect the areas that are most relevant to the business or mission objectives. A good general starter set of impact criteria should, at a minimum, include financial, productivity, business interruption or system availability tolerances, tangible losses (e.g., property, machinery, equipment), physical security, life, health, safety, fines, and legal penalties. As the impact criteria become refined over time, distinguishing the types of risk that would result in immediate direct costs vs. future loss of revenue or liabilities to the organization may be useful to management. An example of an impact criteria table appears in **figure 4.1.**

Figure 4.1—Table of Impact Criteria[16]			
Impact Area	**Low**	**Moderate**	**High**
Reputation	• Reputation is minimally affected; little or no effort or expense is required to recover. Impact is limited to internal/few customers.	• Reputation is damaged, and some effort and expense are required to recover. Impact is limited to local to specific market area.	• Reputation is irrevocably destroyed or damaged. All geographic market areas are affected.
Core customer loss	• Less than 1 percent reduction in customers for one to 30 day(s) due to loss of confidence.	• 1 percent to 5 percent reduction in customers for one to 30 day(s) due to loss of confidence.	• More than 5 percent reduction in customers for one to 30 day(s) due to loss of confidence.
Financial	• Fines less than US $50,000 are levied. • Nonfrivolous lawsuit or lawsuit(s) less than US $250,000 are filed against the organization, or frivolous lawsuit(s) are filed against the enterprise.	• Fines between US $50,000 and US $100,000 are levied. • Nonfrivolous lawsuit(s) greater than US $250,000 but less than US $500,000 are filed against the enterprise.	• Fines greater than US $100,000 are levied. • Nonfrivolous lawsuit or lawsuit(s) greater than US $500,000 are filed against the enterprise.
Productivity/staff hours	• Staff work hours are increased by less than 5 percent for one to 30 day(s). • Less than 25 percent of staff is out for one week or less.	• Staff work hours are increased between 3 percent and 7 percent for 30 to 60 day(s). • More than 25 percent of staff is out for one to three weeks.	• Staff work hours are increased by more than 8 percent for more than 60 days. • More than 25 percent of staff is out for more than three weeks.

Some enterprises assign labels (e.g., very frequent, frequent, infrequent, rare) to the scales. The use of only these labels as a means of expressing frequency is not advisable because they can mean different things for different risk scenarios and consequently can generate confusion. For example, an attempt for network intrusion through the firewall might happen hundreds of times per day, which may be considered "average," whereas the "average" frequency of a hardware failure (e.g., disk crash) might be once every two or three years. So, the word "average" means different frequencies for two different scenarios and, therefore, is not well suited as an objective and unambiguous indicator of frequency.

---

[16] Derived from Software Engineering Institute, Carnegie Mellon University, *http://resources.sei.cmu.edu/library/asset-view.cfm?AssetID=8419*

When analyzing risk scenarios, two properties of each risk scenario need to be assessed: frequency and impact. "Magnitude" and "impact" are often used interchangeably in common risk conversations, although they can carry slightly different nuances. "Magnitude" is a more objective, hard description of how big something is, whereas "impact" is a more subjective description of what this means for enterprises and the consequence it has on business objectives. The technique to describe impact, explained in this section, includes both aspects.

Risk analysis requires the estimation of the frequency of adverse events and their impact (expressed in business terms). Many enterprises use their own developed scales for this purpose, based on one of the techniques described in this section. There is benefit in using the same metrics groups for frequency and impact across the entire (extended) enterprise: to allow better understanding of risk and easier comparisons across the value chains. Ultimately, most risk analysis methods require specific loss data, ranges of actual loss data, estimated ranges of potential data loss, or some combination of these data elements to understand potential impact.

Statistical data may be available in varying quantities and quality, ranging on a continuous scale from almost nonexistent to widely available. When a wide choice of statistical data is available, a quantitative assessment might be the preferred risk assessment method. With very little, incomplete or poor data, or a situation in which the organization wants to learn more about the characteristics of the risk, a wider set of quantitative ranges can be used to faithfully represent the uncertainty arising from sparse data. The better the inputs to either the qualitative or quantitative process, the more reliable the results. Having incomplete, flawed or inaccurate data inputs or erroneous estimates affects the reliability of both quantitative and qualitative methods.

Hybrid risk assessment methods may be applied to situations in between the extremes described here. When less data are available, using simpler measurements and methods may be a good place to start, keeping in mind that the estimates used in the analysis may introduce uncertainty into the analyses. It is also important to communicate to stakeholders the level of confidence in the data or estimates. This provides transparency into the accuracy of the results of the analysis and the analyst's confidence in relying upon those results.

4.1.2 Risk Identification

This phase calls for identifying threats, conditions, areas of concern or known risk to business or mission objectives. Often there is not a proactive risk identification process in an enterprise, which means there is no way to raise an area of concern to the proper level in an organization for decision-making. When working with boards or governance committees, risk is often raised up through the audit process rather than a proactive risk identification process.

In instances of risk, crisis or incident there is most likely someone who knew there was a condition or circumstance that could lead to something bad happening but there was no way for the individual to articulate it or raise a concern that could then be subject to the analysis processes and requisite decision-making techniques.

Organizations are often much better at responding to realized risk than preventing it from occurring when they have a strong risk identification process or capability. Many enterprises today are more comfortable with the risk assessment approach that attempts to determine the frequency of an event or incident occurring (realized risk), which is then combined with an organizationally defined severity scale, impact score or ordinal numerical rating on the impact that would result from the realized risk. The difficulty of using this type of analysis is that probability and impact are important, but they do not present the complete picture of the totality of the risk faced by the organization.

The risk identification process seeks to improve confidence in knowing and understanding the risk that has the potential to impede the enterprise's ability to meet its strategic objectives. The risk identification process can occur in both formal (e.g., brainstorming sessions or workshops) or informal settings (e.g., issues identified in meetings or during "water cooler talk"). When occurring during a brainstorming session, it often starts with a list of things that keep participants up at night, cyberthreats or areas of concern.

The risk identification process seeks to improve confidence in knowing and understanding the risk that has the potential to impede the enterprise's ability to meet its strategic objectives.

The risk practitioner has several possible sources for identification of risk, including:

- Historical or evidence-based methods, such as review of:
  - Audit or incident reports
  - Public media (e.g., newspapers, television)
  - Annual reports and press releases
- Systematic approaches (expert opinion) in which a risk team examines and questions a business process in a systematic manner to determine the potential points of failure, such as:
  - Vulnerability assessments
  - Review of BCPs and DRPs
  - Interviews and workshops with managers, employees, customers, suppliers and auditors
  - Inductive methods (theoretical analysis), in which a team examines a process to determine the possible point of attack or compromise, such as penetration testing

Gathering information from staff is a valuable method of gaining insight into the business from those closest to the processes and most likely to understand their fundamental workings. However, interviews pose certain challenges of which the risk practitioner must be aware. One is that many people want to be seen as essential participants in the mission of the organization, which can lead to exaggeration of their own importance and that of their teams or departments. Another is that people may not fully understand overall business processes or the dependencies between their department and other departments. In certain cases, such as instances of demonstrable negligence or wrongdoing, someone may intentionally provide incorrect information.

Gathering information from staff is a valuable method of gaining insight into the business from those closest to the processes and most likely to understand their fundamental workings.

The risk practitioner can improve the odds of an interview producing useful information by adhering to the following good practices:

- Designate a specific time period and do not exceed that time without mutual agreement.
- When a manager is told that a staff member will be needed for 45 minutes, he/she should not discover that the interview lasted 90 minutes.
- Know as much as possible about the business process in advance of the interview, to reduce time spent on general explanations of core business functions.
- Forge a partnership with the internal audit team responsible for the specific scope of the project or engagement. The benefit of transparency in the risk processes, with audit as a partner, cannot be overstated.
- Before the interview, obtain and review relevant documentation related to the scope of the risk landscape such as process maps, standard operating procedures, the results of impact assessments and network topologies.
- Prepare questions and provide them to interviewees in advance so they can bring any supporting documentation, reports or data that may be necessary.
- Conduct interviews with senior leaders to ensure a thorough understanding of the enterprise, including every aspect of each business operation. Senior leaders may include board members, administrators, critical third-party service providers, customers, suppliers and managers.

- Encourage interviewees to be open about challenges they face and risk that concerns them, as well as any potential missed opportunities or problems associated with their current processes, systems and services/products.

- Avoid setting incorrect expectations regarding confidentiality of interview answers. People may worry about the repercussions of discussing flaws or missed opportunities. Promise confidentiality only if it will actually be maintained.

4.1.3 Risk Assessment and Analysis

"Risk assessment" is often used as a generic term to describe any process used to identify and evaluate risk, whether or not the evaluation of the risk is subject to qualitative or quantitative analysis methods. In its most simple approach, a risk assessment consists of understanding what could go possibly wrong, the likelihood (probability) that a particular event will occur and its potential impact on the enterprise.

Risk assessment is slightly broader than risk analysis and includes the activities of ranking or prioritizing the identified risk according to defined enterprise risk thresholds (based on tolerances), grouping like risk types together for mitigation action, and documenting existing controls that provide mitigation for similar risk types. No matter the assessment or analysis methods used in an enterprise, the risk practitioner can produce a more meaningful result by adhering to the following good practices:

- Perform some assessment (usually qualitative) or analysis (usually quantitative) on the threat, condition or concern to decide on a course of action. Threats, conditions or concerns that are assessed or analyzed to potentially have a significant enough impact on the business, if realized, may also be evaluated for probability of occurrence. A common analysis method in this phase of the process is to use the MFL or the PML quantitative analysis methods. These methods are able to help management understand the total financial impact on the enterprise should the risk be realized. MFL and PML are best coupled with a thorough set of relevant risk scenarios with well-stated assumptions.

- If a risk is identified, enter it into a list, sometimes called a risk register, and determine the next step in analysis or response. This step in the risk management process often needs further analysis of the risk factors to determine an effective course of action or cost-justification for a plan of remediation. A recommended decision analysis process is to use Monte Carlo[17] modeling and simulation analysis to rank-stack, or prioritize, the list of risk concerns in a risk register for appropriate responses. Monte Carlo simulations, such as those used in the Factor Analysis of Information Risk[18] (FAIR™) method, view risk as a function of likelihood (frequency of something happening) and impact. The full analytic model of FAIR enables Monte Carlo analyses; however, it can be used to statically assess best- and worst-case outcomes of scenarios to enable quick triaging. The point here is that likelihood and impact are not the whole picture when analyzing risk. Unlikely events occur all too often, and many likely events never materialize.

- Annual loss expectancy (ALE) is sometimes used to spread the probable losses over some period of time depending on the accuracy of the data used in the modeling. Using the ALE method coupled with a Monte Carlo analysis may not be helpful for making risk transfer decisions. ALE spreads the loss over a time horizon that may distort or minimize the actual financial losses that would be realized if the risk was realized. For example, assume that using a Monte Carlo simulation for a given scenario determines the impact of a specific cyberrisk to be US $350 million should it occur. The probability of the risk materializing is once every 10 years. Combining the result of the Monte Carlo model analysis of a US $350 million impact with ALE of the probability of this risk materializing once every 10 years spreads the US $350 million impact over a period of 10 years. The analysis of this specific cyberrisk results in placing it on the risk register as a potential US $35 million loss, which could be in tolerance for the organization. However, when that specific cyberrisk actually materializes in a given year, the enterprise has a loss event of US $350 million, not US $35 million. The organization might have

17 Brownlee, J.; "A Gentle Introduction to Monte Carlo Sampling for Probability," Machine Learning Mastery, 4 November 2019, *https://machinelearningmastery.com/monte-carlo-sampling-for-probability/*
18 The Open Group, Open FAIR™standards, *https://publications.opengroup.org/standards/open-fair-standards*

responded with different mitigation or risk transfer options had it known of the full potential impact (MFL/PML) of the loss before it materialized.

Several methods for risk analysis exist, ranging from high-level and mostly qualitative to very detailed and/or quantitative, with hybrid methods in between. Many forms may be needed at different stages of the risk management process. For example, qualitative analysis tends to be better at the initial risk assessment stage to perform a quick triage of identified risk, and quantitative analysis can then provide additional rigor and accuracy for the selected risk types or areas that need further analysis. There are many good sources of data to support risk analysis, and there are a number of sources where these data can be obtained internally, including colleagues from other disciplines that may be collecting similar data.

The enterprise's culture, resources, skills and knowledge of I&T risk management, environment, risk appetite, and its existing approach to ERM will determine which methodology is used.

4.1.4 Qualitative and Quantitative Approaches

A qualitative risk assessment approach uses expert opinions to estimate the frequency and business impact of adverse events. The frequency and the impact are estimated using qualitative descriptions such as high, medium or low. These labels can vary depending on the circumstances and different environments. To further explain this point, it is not generally useful to group risk using a qualitative descriptor because there is no way to know which "high" risk is higher than another "high" risk in the same bucket. Alternatively, there is no way to know if putting together a group of "low" risk is equivalent to one or more "medium" or one "high" risk. Qualitative methods also do not support risk aggregation in a meaningful way when attempting to look at risk across the enterprise.

Qualitative methods are effectively used to identify like or unlike characteristics for grouping purposes and may assist in efficiencies in applying controls or mitigation techniques to respond to risk that would be effective against common risk types. For example, it is possible to reduce uncertainty by grouping all the "high" risk and further evaluating if any of the risk in that category may be mitigated or responded to in an efficient manner that adds value to the enterprise. This may result in improved procedures, streamlined compliance objectives or optimized control objectives. Qualitative approaches are not meant to be a measurement method. Measurement of risk should assist an enterprise with decision-making about the uncertainty of a risk scenario materializing.

Qualitative methods are effectively used to identify like or unlike characteristics for grouping purposes and may assist in efficiencies in applying controls or mitigation techniques to respond to risk that would be effective against common risk types.

When quantitative values (e.g., estimated ranges, financial, time) are used to define qualitative values, or when only quantitative values are used, it is a quantitative analysis. For example, many organizations currently using a measurement scale of high, medium and low will arbitrarily assign the risk rating scale to the numbers 1 (high), 2 (medium) or 3 (low) to rank-stack a list of risk. Using a numeric scale to plot identified risk on a chart does not equal quantitative risk analysis. The essence of quantitative risk assessment is to derive the frequency and impact of risk scenarios, based on measurement, statistical methods and data.

Analysis based on individual opinions or estimated data may be insufficient to make better decisions and may not provide value to the enterprise. How certain can one be about the results of risk assessment if the inputs to the results are flawed, biased or incorrect? There is still the question of uncertainty. Some advanced methods exist to increase reliability of risk assessments, but these require some statistical knowledge and skills. Currently, calibration of a risk practitioner is not based on statistical methods, yet it does improve the estimating process and the reliability of quantitative inputs when data are sparse. Calibration is the fine-tuning of risk practitioners' ability to improve their estimates over time, recognize their own and other biases, and decrease subjectivity in the analysis phase. Quantitative risk analysis as currently applied in practice is very much in the early stages of maturity in the I&T field and investing in advanced quantitative methods does not always result in the envisioned value or return on the investment.

The different methods—quantitative and qualitative—have some common limitations:

- No method is fully objective, and the results of the assessments and analysis may be subject to personal bias, knowledge and skills of the assessor, and available data used in the evaluation.
- Quantitative-only evaluations are subject to creating overconfidence in complex models based on insufficient data.
- Qualitative-only evaluations are also subject to unreliable results because of the oversimplified nature of using an ordinal or taxonomy-based measurement scale.

4.1.5 Risk Maps (Heatmaps)

As stated earlier in this document, risk maps—or heatmaps—have become the common method to visualize the risk context and scope. On these maps, risk is plotted on a two-dimensional diagram, with frequency and impact being the two dimensions. A sample risk map is shown in **figure 4.2.**

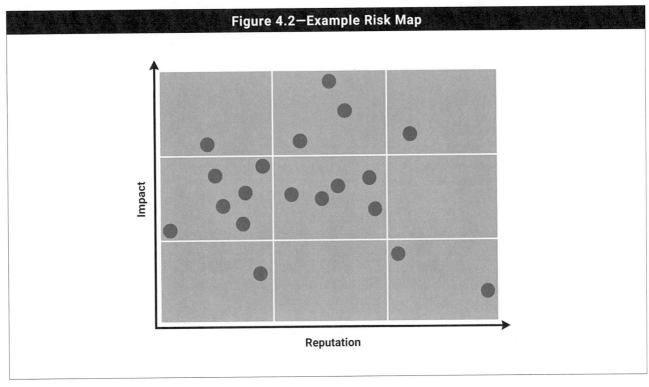

Figure 4.2—Example Risk Map

The risk map becomes more useful when it is combined with the different zones of risk appetite (see chapter 3). Different risk appetite bands of significance are defined in the risk map using colored zones, leading to the example in **figure 4.3**. This version of the risk map immediately identifies the risk that is truly unacceptable and requires an immediate response, as defined by the enterprise risk appetite statements.

At the other end of the spectrum, the risk map could also allow for the identification of opportunities for relaxing controls or taking on more risk, as represented by the blue zone in **figure 4.3.**

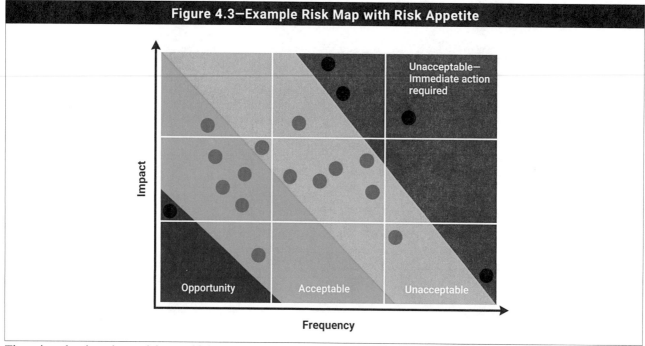

Figure 4.3—Example Risk Map with Risk Appetite

There is value in using a risk map if the underlying measurement criteria and impact scales are well defined and the language is well understood by the organization. As the risk management process and risk decision-making mature, there is often a divergence away from calculating the probability of a risk occurring and increased focus on the financial impact the realized risk could have on the enterprise. This occurs when the board and management have a requirement to quantify risk in financial terms; the most common techniques for that are to shift the focus to the calculation of the PML or the MFL. Using PML or MFL allows the risk to be analyzed with regard to the financial impact on the organization should the risk be realized. These techniques are used for ensuring the enterprise has the appropriate financial protections in place should a risk materialize.

As the risk management process and risk decision-making mature, there is often a divergence away from calculating the probability of a risk occurring and increased focus on the financial impact the realized risk could have on the enterprise.

A final consideration when estimating impact during risk assessment and analysis is that the best way to obtain reliable and accepted estimations is to involve all stakeholders in scenario analysis exercises. This can be done through separate assessments or through workshops, followed by group discussion to achieve consensus. The staff performing the work are great sources of scenarios that are plausible, or they may have information on near misses that have occurred. The assessment confidence level is the trust in the data sources or the calibrated estimates used in the analysis, and the advantages of communicating that level to audiences cannot be overstated. This can be especially important when less sophisticated ways of rating or visualizing risk are used to ensure decision-makers have a sense for how much to trust the assessment results.

4.1.6 Risk Register

A list of risk areas that have been identified, analyzed and prioritized is called a **risk register.** A risk register is the list of potential adverse events that have been identified and analyzed to understand the potential impact should the risk be realized. The risk register is not a list of control deficiencies or missing software patches from a server. If there is no uncertainty, the control deficiency is an issue or problem, not a risk. Issue or problem management is beyond the scope of this publication but could be considered as part of setting up the risk function. The risk register can be seen as an extension of the risk map (see **figure 3.3**), providing detailed information on each identified risk, including:

- Risk identification
- Risk owner
- Details of the risk scenario
- Organization or business unit
- Date of risk identification
- Source of risk (if known)
- Risk owner/point of contact
- Risk title
- Risk statement
- Heat (map) score (qualitative)
- Information on detailed analysis results or scores (quantitative)
- Detailed information on risk response
- Current risk response status
- Information on controls (if applicable)
- Primary risk category
- Secondary risk category (if using)
- Remediation or disposition status
- Current status date
- Follow-up date
- Comments

There is no new information in the risk register not covered in the previous sections. A risk register is just a convenient technique to store and maintain all collected information in a useful format for all stakeholders. Often when an enterprise decides to purchase a commercial tool for managing risk it includes an out-of-the-box risk register template that can be customized for the unique organization.

Page intentionally left blank

Chapter 5
Risk Scenarios Defined

5.1 Introduction

Risk scenarios facilitate communication in risk management by constructing a narrative that can inspire people to take action. The use of risk scenarios can enhance the risk management effort by helping the risk team to understand and explain risk to the business process owners and other stakeholders. Additionally, a well-developed scenario provides a realistic and practical view of risk that is more aligned with business objectives, historical events and emerging threats envisioned by the organization than would be found by consulting a broadly applicable standard or catalog of controls. These benefits make risk scenarios valuable as means of gathering and framing information used in subsequent steps in the risk management process.

One of the challenges for I&T risk management is to identify the important and relevant risk among all that can possibly go wrong with I&T or in relation to I&T, given the pervasive presence of I&T and the business's dependence on it. One of the techniques to overcome this challenge is the development and use of risk scenarios. It is a core approach to bringing realism, insight, organizational engagement, improved analysis and structure to the complex matter of I&T risk. Once these scenarios are developed, they are used during the risk analysis, in which frequency and business impacts are estimated.

Risk scenarios can be derived via two different mechanisms:

- A **top-down approach**, in which mission strategy and business objectives are used as the basis for identification and analysis of the risk that is plausible and relevant to meeting the desired outcomes. If the impact criteria are well aligned with the real value drivers of the enterprise, relevant risk scenarios will be developed.

- A **bottom-up approach**, usually beginning with assets, systems or applications deemed important to the enterprise and followed by use of a list of threats or generic loss scenarios to define a set of more concrete and customized scenarios applied to the individual enterprise situation. The bottom-up approach is commonly used in cyberthreat and vulnerability assessments but it may limit or overlook true business impact if it is not combined with a thorough consideration of the top-down approach detailed above.

The approaches are complementary and should be used simultaneously. This is when a **taxonomy** of risk may be useful. A risk taxonomy provides a scheme for classifying sources and categories of risk. The path from a cyberthreat or area of concern to a risk requires that the statement of risk be decomposed into components that are actionable. A risk taxonomy provides a common language for discussing and communicating risk to stakeholders. Risk scenarios must be relevant and linked to real business or mission risk.

Once the set of risk scenarios is defined, it can be used for risk analysis, in which the frequency and impact of the scenario are assessed. An important component of this assessment are risk factors. Risk factors are those factors that influence the frequency and/or business or mission impact of risk scenarios; they can be of different natures and classified as shown in **figure 5.1**:

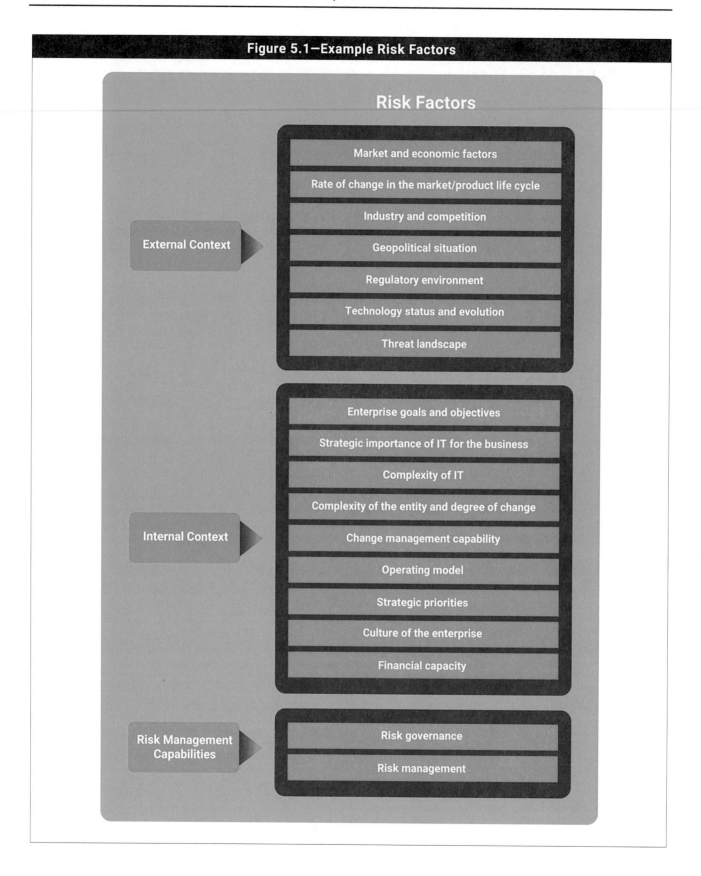

Figure 5.1—Example Risk Factors

Contextual factors can be internal or external to the enterprise, with the main difference being the degree of control that an organization has over them:

- External contextual factors are, to a large extent, outside the control of the enterprise. For example, a mature risk management approach will spend less time on performing analysis on the probability of a threat occurring and more time on developing capabilities that would more quickly detect and contain the threat if it materializes.

- Internal contextual factors are, to a large extent, under the control of the enterprise, although they may not always be easy to change. For example, poor user awareness training practices are often a root cause in a risk that is realized and results in an incident.

The following structure for risk scenarios is proposed, based on a review of current risk methodologies (Risk IT, COBIT 2019, FAIR, Committee of Sponsoring Organizations of the Treadway Commission [COSO] Enterprise Risk Management [ERM] 2017). The structure indicates all elements that should be described to have a full, analyzable risk scenario.

The main difference between the definition of a risk scenario presented in the Risk IT framework and the updated COBIT risk scenarios structure is the addition of the effect component. This component is self-explanatory: It describes the effect of the threat event materializing.

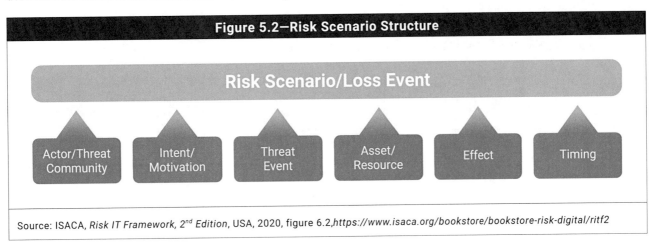

Figure 5.2—Risk Scenario Structure

Source: ISACA, *Risk IT Framework, 2nd Edition*, USA, 2020, figure 6.2, *https://www.isaca.org/bookstore/bookstore-risk-digital/ritf2*

The different components of the risk scenario structure are shown in **figure 5.2** and described as follows:

- **Actor/Threat community**—Who generates the threat that exploits a vulnerability? Actors can be internal or external, and they can be human or nonhuman. Not every type of threat requires an actor; for example, failures or natural causes may constitute threats. Actors are categorized as follows:
  - Internal actors are within the enterprise, such as staff and contractors.
  - External actors include outsiders, competitors, regulators and the market.

- **Intent/Motivation** (the nature of the event)—Is it malicious? If not, is it accidental or is it a failure of a well-defined process? Is it a natural event?

- **Threat event**—Is it disclosure of confidential information, interruption of a system or a project, theft or destruction? Action also includes ineffective design of systems, processes, etc.; inappropriate use; changes in rules and regulation that will materially impact a system; or ineffective execution of processes, such as change management procedures, acquisition procedures and project prioritization processes.

- **Asset/resource**—An asset is any item that is of value to the enterprise or exposes the enterprise to liability that can be affected by the event and lead to business impact. A resource is anything that helps to achieve IT goals. Assets and resources can be identical; for example, IT hardware is an important resource because all IT applications use it, and at the same time it is an asset because it has a certain value to the enterprise.

Assets can be critical or not, as indicated by the difference between a client-facing website of a major bank and the website of the local garage or the intranet of the software development group. Critical resources will probably attract a greater number of attacks or greater attention on failure; hence, the frequency of related scenarios will probably be higher. It takes skill, experience and thorough understanding of dependencies to understand the difference between a critical asset and a noncritical asset.

- Assets/resources include:
  - People and skills
  - Organizational structures
  - Physical infrastructure, facilities, equipment, etc.
  - IT infrastructure, including computing hardware, network infrastructure, middleware
  - Other enterprise architecture components, including information, applications
- **Effect**—This component reflects the effect of the threat scenario, typically a negative effect.
- **Time**—This is the dimension in which the following could be described, if they are relevant to the scenario:
  - The duration of the event, such as an extended outage of a service or data center
  - The timing (Does the event occur at a critical moment?)
  - Detection (Is detection immediate or not?)
  - Time lag between the event and impact (Is there an immediate impact, such as network failure or immediate downtime, or a delayed impact, such as wrong IT architecture with accumulated high costs over a time span of several years?)

It is important to stay aware of the differences between loss events, threat events and vulnerability events. When a risk scenario materializes, a loss event occurs. The loss event has been triggered by a threat event (threat type plus event). The frequency of the threat event leading to a loss event is influenced by the risk factors or vulnerability. Vulnerability is usually a state and can be increased/decreased by vulnerability events, for example, the weakening of controls or the threat strength. **These three types of events should not be combined into one big "risk list."**

Chapter 6
Guidance for Building Risk Scenarios

6.1 Developing Risk Scenarios

In practice, the following approach for building IT risk scenarios is suggested:

- Use the list of example generic risk scenarios presented in **figure 6.2** to define a manageable set of tailored risk scenarios for the enterprise. To determine a manageable set of scenarios an enterprise might begin by considering commonly occurring scenarios in its industry or product area, scenarios representing threat sources that are increasing in number or severity, and scenarios that involve legal and regulatory requirements applicable to the business. Another approach might be to identify high-risk business units and assess one or two high-risk operating processes within each, including the IT components that enable that process. Some less common situations should also be included in the scenarios.

- Perform a validation against the business objectives of the entity. Do the selected risk scenarios address potential impacts on achievement of business objectives of the entity, in support of the overall enterprise's business objectives?

- Refine the selected scenarios based on this validation; detail them to a level in line with the criticality of the entity.

- Reduce the number of scenarios to a manageable set. "Manageable" does not signify a fixed number but the number should be in line with the overall importance (size) and criticality of the unit. There is no general rule, but if scenarios are reasonably and realistically scoped, the enterprise should expect to develop at least a few dozen scenarios.

- Keep all scenarios in a list so they can be reevaluated in the next iteration and included for detailed analysis if they have become relevant at that time.

- Include in the scenarios an unspecified event, for example, an incident not covered by other scenarios.

- Consider evaluating scenarios that have a chance of occurring simultaneously. This is frequently referred to as stress testing and actually entails combining multiple scenarios and understanding what the extra impact would be of them occurring together.

- Base the scenario analysis not only on past experience and known current events but also possible future circumstances.

Once the set of risk scenarios is defined, it can be used for risk analysis, in which frequency and impact of the scenario are assessed. An important component to be considered during such assessments are the risk factors see **figure 5.1**.

6.1.1 Main Issues When Developing and Using Risk Scenarios

The use of scenarios is key to risk management, and the technique is applicable to any enterprise. Each enterprise needs to build a set of scenarios (containing the components described in chapter 5) as a starting point to conduct its risk analysis.

Building a complete set of scenarios means—in theory—that each possible value of every component should be combined. Each combination should then be assessed for relevance and realism and, if found to be relevant, entered into the risk register. In practice, this is not possible; very quickly, an unfeasible number of different risk scenarios can be generated. The number of scenarios to be developed and analyzed should be kept to a relatively small number of relevant risk scenarios to remain manageable and meaningful.

Figure 6.1 shows some of the main areas of focus/issues to address when using the risk scenario technique.

Figure 6.1—Risk Scenario Technique Main Issues/Attention Points	
Issue/Attention Point	**Summary Guidance**
Maintain currency of risk scenarios and risk factors.	Risk factors and the enterprise change over time; hence, scenarios will change over time, over the course of a project or over the evolution of technology.
	For example, it is essential for the risk function to develop a review schedule and the chief information officer (CIO) to work with the business lines to review and update scenarios for relevance and importance. Frequency of this exercise depends on the overall risk profile of the enterprise and should be done at least on an annual basis, or when important changes occur.
Use generic risk scenarios as a starting point and build more detail where and when required.	One technique of keeping the number of scenarios manageable is to propagate a standard set of generic risk scenarios through the enterprise and develop more detailed and relevant scenarios when required and warranted by the risk profile only at lower (entity) levels. The assumptions made when grouping or generalizing should be well understood by all and adequately documented because they may hide certain scenarios or be confusing when looking at risk response.
	For example, if "insider threat" is not well defined within a scenario, it may not be clear whether this threat includes privileged and nonprivileged insiders. The differences between these aspects of a scenario can be critical in trying to understand the frequency and impact of events, as well as mitigation opportunities.
The number of scenarios should be representative and reflect business reality and complexity.	Risk management helps to deal with the enormous complexity of today's IT environments by prioritizing potential action according to its value in reducing risk. Risk management is about reducing complexity, not generating it, hence another plea for working with a manageable number of risk scenarios. However, the retained number of scenarios still needs to accurately reflect business reality and complexity.
Risk taxonomy should reflect business reality and complexity.	There should be a sufficient number of risk scenario scales reflecting the complexity of the enterprise and the extent of exposures to which the enterprise is subject.
	Potential scales might be a low-medium-high ranking or a numeric scale that scores risk importance from 0 to 5. Scales should be aligned throughout the enterprise to ensure consistent scoring.
Use a generic risk scenario structure to simplify risk reporting.	Similarly, for risk reporting purposes, entities should not report on all specific and detailed scenarios but can do so by using the generic risk scenario structure.
	For example, an entity may have taken generic risk scenario 2 in **figure 6.2** (project quality), translated it into five scenarios for its major projects, subsequently conducted a risk analysis for each of the scenarios, then aggregated or summarized the results and reported back using the generic risk scenario header "project quality."
Ensure the enterprise has adequate people and skills requirements for developing relevant risk scenarios.	Developing a manageable and relevant set of risk scenarios requires: • Expertise and experience to avoid overlooking relevant scenarios and being drawn into highly unrealistic or irrelevant scenarios. While the avoidance of scenarios that are unrealistic or irrelevant is important in properly utilizing limited resources, some attention should be paid to situations that are highly infrequent and unpredictable, but which could have a cataclysmic impact on the enterprise. • A thorough understanding of the environment. This includes the IT environment (e.g., infrastructure, applications, dependencies among applications, infrastructure components), the overall business environment, and an understanding of how and which IT environments support the business environment to understand the business impact. • The intervention and common views of all parties involved—senior management, which has the decision power; business management, which has the best view on business impact; IT, which has the understanding of what can go wrong with IT; and risk management, which can moderate and structure the debate among the other parties. • The process of developing scenarios usually benefits from a brainstorming/workshop approach, which usually requires performance of a high-level assessment to reduce the number of scenarios to a manageable, but relevant and representative, number.

Figure 6.1—Risk Scenario Technique Main Issues/Attention Points *(cont.)*	
Issue/Attention Point	**Summary Guidance**
Use the risk scenario building process to obtain buy-in.	Scenario analysis is not just an analytical exercise involving risk analysts. A significant additional benefit of scenario analysis is achieving organizational buy-in from enterprise entities and business lines, risk management, IT, finance, compliance, and other parties. Gaining this buy-in is the reason why scenario analysis should be a carefully facilitated process.
Involve the first line of defense in the scenario building process.	In addition to coordinating with management, it is recommended that selected members of the staff who are familiar with the detailed operations be included in discussions, where appropriate. Staff whose daily work is in the detailed operations are often more familiar with vulnerabilities in technology and processes that can be exploited.
Do not focus only on rare and extreme scenarios.	When developing scenarios, the focus should not be placed only on worst-case events because they rarely materialize; less severe incidents happen more often.
Deduce complex scenarios from simple scenarios by showing impact and dependencies.	Simple scenarios, once developed, should be further fine-tuned into more complex scenarios, showing cascading and/or coincidental impacts and reflecting dependencies. For example: • A scenario of having a major hardware failure can be combined with the scenario of a failed DRP. • A scenario of a major software failure can trigger database corruption and, in combination with poor data management backups, can lead to serious consequences, or at least consequences of a different magnitude than a software failure alone. • A scenario of a major external event can lead to a scenario of internal apathy.
Consider systemic and contagious risk.	Attention should be paid to systemic and/or contagious risk scenarios: • **Systemic**—Something happens with an important business partner, affecting a large group of enterprises within an area or industry. An example would be a nationwide air traffic control system that goes down for an extended period of time, such as six hours, affecting air traffic on a very large scale. • **Contagious**—Events that happen at several of the enterprise's business partners within a very short time frame. An example would be a clearinghouse that can be fully prepared for any sort of emergency by having very sophisticated disaster recovery measures in place, but when a catastrophe happens, finds that no transactions are sent by its providers and hence is temporarily out of business.
Use scenario building to increase awareness for risk detection.	Scenario development also helps to address the issue of detectability, moving away from a situation where an enterprise does not know what it does not know. The collaborative approach for scenario development assists in identifying risk to which the enterprise, until then, would not have realized it was subject (and hence would never have thought of putting in place any countermeasures). After the full set of risk items is identified during scenario generation, risk analysis assesses frequency and impact of the scenarios. Questions to be asked include: • Will the enterprise ever detect that the risk scenario has materialized? • Will the enterprise notice something has gone wrong so it can react appropriately? Generating scenarios and creatively thinking of what can go wrong will automatically raise and, hopefully, cause responses to, the question of detectability. Detectability of scenarios includes two steps: visibility and recognition. The enterprise must be in a position to observe anything going wrong, and it needs the capability to recognize an observed event as something wrong.

Figure 6.2 shows example generic and specific risk scenarios. It is intended to help the risk practitioner understand and predict potential situations in order to holistically develop the appropriate risk response.

Figure 6.2—Example Generic and Specific Risk Scenarios		
Risk Scenario Category	Example Generic Risk Scenarios	Example Specific Risk Scenarios
1 IT investment decision making, portfolio definition and maintenance	A Programs selected for implementation are misaligned with corporate strategy and priorities.	• Programs selected for implementation are misaligned with corporate strategy and priorities, resulting in an inappropriate new customer relationship management (CRM) application that fails to support the customer service strategy.
	B IT-related investments are not supporting the digital strategy of the enterprise.	• The enterprise has a misaligned investment in a new accounting system that limits the investment funds available for the multichannel customer-facing application.
	C Wrong software (cost, performance, features, compatibility, redundancy, etc.) is selected for acquisition and implementation.	• Incompatible software for older production machines is implemented to replace legacy systems, resulting in loss of production and business.
	D Wrong infrastructure (cost, performance, features, compatibility, etc.) is selected for implementation.	• An investment in on-premise infrastructure is made despite lacking requisite capabilities to successfully implement and support, resulting in increased costs and service failures. • An investment in cloud infrastructure is made that results in limited local availability necessary to support business goals, resulting in service availability failures.
	E Duplication or important overlaps among different investment initiatives exist.	• Two large departments in the enterprise are heavily investing in new customer-facing applications but both are developing their own (different) customer database, resulting in increased organizational costs.
	F New investment programs create long-term incompatibility with the enterprise architecture.	• Programs selected for implementation are misaligned with the enterprise architecture, resulting in an inappropriate new accounting application that fails to integrate with the current application landscape, resulting in increased costs and service failures.
	G Competing resources are allocated and managed inefficiently and are misaligned to business priorities.	• Technical experts in critical technologies are assigned to work on less critical projects instead of high-priority and key projects, resulting in product and service delivery failures.
	H Shadow IT and Software as a Service (SAAS) solutions are deployed from departmental budgets, generating longer-term architectural problems and excessive overall IT cost.	• The marketing department acquires its own SAAS-based customer management system, resulting in expensive additional developments of adequate interfaces with legacy systems.
	I Inadequate consideration is given to future requirements or demands.	• The criteria for IT investment selection do not include future user requirements, leading to incomplete solutions or solutions not aligned with business strategy.
	J Initiative economics are not considered during investment decisions.	• Project economics are not adequately used for investment selection, leading to high-return projects being discarded or negative return projects not abandoned.

Figure 6.2—Example Generic and Specific Risk Scenarios (cont.)

Risk Scenario Category		Example Generic Risk Scenarios	Example Specific Risk Scenarios
2 Program and projects life cycle management	A	Failing projects (cost explosion, excessive delays, scope creep, changed business priorities) are not terminated by senior management.	• A long-running and ambitious project to build a new human resources (HR) system has not yet delivered any tangible results in its attempt to integrate payroll management, time management and performance management. The enterprise, despite the lack of results, keeps (under)funding the project for the next period.
	B	IT projects exceed planned budgets.	• A major project for a new enterprise resource planning (ERP) implementation approaches a 100 percent budget overrun.
	C	IT projects do not meet business requirements.	• The new CRM system is performing unacceptably slowly and is unstable because of many crashes, resulting in an availability rate of less than 90 percent. • The new CRM system's implementation of privacy regulations fails for 50 percent of the test cases during acceptance testing, creating a noncompliance exposure. • The new CRM system's functionality and user interface (UI) are not accepted by the intended users, due to not obtaining the active involvement throughout the program/project life cycle of all stakeholders (including the sponsor). • Because of incomplete business requirements analysis and/or functional design, the new HR system is not in line with users' expectations. • Because of an ineffective testing process, the new ERP system contains many bugs and impacts productivity.
	D	IT projects are delivered late.	• The enterprise's new website, including the online shop, is delivered six months behind schedule, creating a missed market opportunity.
	E	New application software is not adopted by users.	• Incorrect requirements analysis results in users' failure to adopt the new CRM application and a loss of efficiency.
	F	Immature software (early adopters, bugs, etc.) is implemented.	• Business application experiences service disruption due to software defects.
	G	Security requirements are inadequately assessed.	• The lack of consideration for security requirements leads to an insecure solution for a new remote access application, requiring extensive rework and delays.
	H	Program/project scope is not well managed.	• Frequent and late program scope changes forced by the system owner upon the developer lead to unstable and inconsistent solutions, causing extensive rework and delays.
	I	Inadequate KPIs exist to measure program purpose achievement.	• The enterprise is unable to manage projects correctly because it is unaware of its true state, leading to late/wrong project decisions, delays and extensive overruns.

	Risk Scenario Category		Example Generic Risk Scenarios	Example Specific Risk Scenarios
3	IT cost and oversight	A	There is extensive dependency on and use of user-created and ad hoc solutions.	• Extensive dependency on the use of end-user computing for important information needs leads to security deficiencies. • Extensive dependency on the use of end-user computing for important information needs leads to inaccurate and unreliable data. • Extensive dependency on ad hoc solutions for important information needs leads to inefficient use of resources and additional cost.
		B	Change management over ad hoc solutions is inadequate.	• Inadequate change management and quality control over ad hoc user solutions leads to wrong computing results and business decisions (e.g., business case outcomes are wildly inaccurate).
		C	Cost and ineffectiveness are related to I&T-related purchases outside of the I&T procurement process.	• Different departments are acquiring their own office automation solutions. • Different departments are entering into agreements with different outsourcing vendors or service providers.
		D	Inadequate requirements lead to ineffective service level agreements (SLAs).	• Relevant stakeholders did not participate in the requirements phase, resulting in incomplete requirements and incomplete SLAs.
		E	There is a lack of funding for I&T-related investments.	• Insufficient funding for necessary security updates leads to failure of IT systems through cyberattacks. • Lack of funding for new I&T innovations results in a slower go-to-market response and a loss of competitive advantage.
4	IT expertise, skills and behavior	A	There is a lack or mismatch of IT-related skills within IT (e.g., due to new technologies or working methods).	• Current IT staff do not have the required skills for new technologies (e.g., blockchain), resulting in longer learning curves and delays in projects. • Current IT operations and development staff are not trained in DevOps tools and ways of working, resulting in less-than-anticipated benefits of the DevOps approach.
		B	A lack of business understanding by IT staff affects service delivery/projects quality.	• A lack of business understanding by IT staff affects the quality of the new ERP system due to deficient implementation of user requirements and implementation of unnecessary features.
		C	The enterprise is unable to recruit and retain IT staff.	• The HR department chronically fails to recruit sufficiently skilled information security experts, leading to increased exposure and increased costs for external experts. • There is an insufficient return on investment regarding training due to early departure of trained IT staff (e.g., individuals with advanced business degrees).
		D	Individuals with unsuitable profiles are recruited because of lack of due diligence in the recruitment process.	• The HR department recruits employees with false credentials, which results in damage to the brand in the long run. • A failure to perform the necessary due diligence on a new hire resulted in insider hacking of confidential data.

Figure 6.2—Example Generic and Specific Risk Scenarios *(cont.)*

Risk Scenario Category		Example Generic Risk Scenarios	Example Specific Risk Scenarios
		Figure 6.2—Example Generic and Specific Risk Scenarios *(cont.)*	
4	**IT expertise, skills and behavior** *(cont.)*	E	I&T training is lacking.	• The lack of I&T training leads to the departure of IT staff. • The lack of I&T-related training leads to increased quality problems of delivered services. • The enterprise is unable to update the I&T skills to the proper level through training.
		F	There is an overreliance on key staff for I&T services.	• Critical in-house knowledge is lost when key personnel leave the organization.
		G	No formal staff performance review exists.	• The lack of a formal performance review of IT development staff leads to subpar software quality and delayed delivery time of new applications.
		H	Duties are ineffectively segregated.	• Ineffective segregation of duties for the purchasing application allows for fraudulent transactions.
		I	Communication challenges exist between IT and users.	• Difficult relationships between IT and users lead to poorly understood user requests and deficient solutions.
5	**Enterprise/IT architecture**	A	Enterprise architecture is complex and inflexible, obstructing further evolution and expansion, leading to missed business opportunities.	• A new customer relationship management application cannot be implemented or has to be reduced in functionality because communication with other internal systems is too slow because of very slow interfaces.
		B	The enterprise fails to adopt and exploit new infrastructure or abandon obsolete infrastructure on a timely basis.	• Abandoned infrastructure cabling reduces the efficiency of heating, venting and air conditioning (HVAC) systems.
		C	The enterprise fails to adopt existing enterprise/IT architecture when designing and implementing new technology solutions.	• The marketing department decides to use a SAAS solution without consulting the enterprise architects, resulting in a solution that is incompatible with current internal systems, requiring the expensive build of additional interfacing software.
		D	The enterprise fails to adopt and exploit new software (functionality, optimization, etc.) or abandon obsolete applications on a timely basis.	• The crash of an older, unsupported accounting application results in high profit loss. • The failure to abandon obsolete applications takes up unnecessary time and resources (i.e., money and people).
		E	Undocumented enterprise architecture leads to inefficiencies and duplications.	• The marketing department decides to use a SAAS solution without consulting the enterprise architects, resulting in a solution that is incompatible with current internal systems, requiring the expensive build of additional interfacing software.
		F	There is an excessive number of exceptions on enterprise architecture standards.	• Over the previous year, almost 75 percent of the new projects have requested (and been granted) exceptions to the enterprise architecture standards. This renders the standards in practice irrelevant and will create inconsistent solutions and excessive cost going forward.

Figure 6.2—Example Generic and Specific Risk Scenarios *(cont.)*			
Risk Scenario Category		Example Generic Risk Scenarios	Example Specific Risk Scenarios
6 IT operations	A	Errors are made by IT staff (during backup, during upgrades of systems, during maintenance of systems, etc.).	• IT operations staff (internal or at the service provider) enters a wrong command during an upgrade of a system, leaving the system vulnerable, from a security point of view.
	B	Incorrect information is input by IT staff or system users.	• The system manager of a critical server enters incorrect network information into the system, crippling the communication speed of the system to unacceptable levels.
	C	Backup/recovery management is deficient and includes mislabeling or misplacing backup media.	• Deficient backup/recovery management is causing data loss when backups cannot be restored after a hardware incident. • Deficient backup/recovery management is causing data loss when backups cannot be restored after a ransomware incident. • Backup status is not properly monitored and/or backup issues are not being resolved in a timely manner, leading to potential data loss when a restore would be needed.
	D	Patch/vulnerability management is inadequate.	• Inadequate patch and vulnerability management leaves systems vulnerable to attacks or crashes, reducing service levels.
	E	Performance and operations monitoring is inadequate.	• Inadequate systems or application monitoring can cause system degradation to remain unnoticed for too long, leading to service interruptions.
	F	Facilities' resilience is inadequate.	• Power interruptions or surges without an adequate Unterruptible Power Supply system damage computing equipment and can cause service interruptions or data loss.
7 User access rights management	A	Software is tampered with.	• An IT operations manager in a DevOps environment makes a number of software changes he believes will make a new application run smoother inadverntly causing an error or a system performance issue.
	B	Software is intentionally modified or manipulated, leading to wrong data.	• The modification of database source code by a developer with wrong intentions results in wrong data.
	C	The security configuration is intentionally modified or manipulated.	• Tampering with the security configuration of network equipment creates important vulnerabilities, allowing attacks to be successful.
	D	Software is intentionally modified or manipulated, leading to fraudulent actions.	• A software engineer makes unauthorized changes to a payment system so he/she can make unauthorized and fraudulent payments.
	E	Software is unintentionally modified, leading to inaccurate results.	• Unintentional false modification of database source code by an unaware developer results in loss of production and data.
	F	Unintentional errors are made in configuration and change management.	• Change management or configuration errors may cause system outages, leading to a service interruption of customer-facing applications.

Figure 6.2—Example Generic and Specific Risk Scenarios *(cont.)*		
Risk Scenario Category	Example Generic Risk Scenarios	Example Specific Risk Scenarios
7 User access rights management *(cont.)*	G Unintentional communications are issued.	• An email with sensitive personal information is sent to the wrong recipients, leading to breach of privacy claims.
	H Access rights from prior roles are abused to access IT infrastructure.	• A lack of internal control of assigned roles leads to improper access rights of users. Financial data are accessible and stolen by employees.
	I Authorized users intentionally or accidentally conduct unauthorized actions.	• A privileged user is accessing private personal information from clients for no valid business reason, exposing the enterprise to a privacy violation.
	J Access rights or passwords are shared with others.	• Authorized users (privileged or not) share their credentials with other users, allowing these other users actions to which they are not entitled and obstructing nonrepudiation.
	K Privileged/emergency account management is inadequate.	• Deficient management of privileged accounts results in an excessive number of users with privileges, increasing the likelihood of unauthorized actions and reducing nonrepudiation.
	L User access provisioning is ineffective.	• Policy dispensation processes during a disaster are ineffective. (During a disaster, some existing controls may not be feasible. Thus, there should be a dispensation process to assess the risk and track for the dispensation.)
	M The access matrix is not defined or implemented.	• A poorly designed or implemented access matrix may result in users with (typically) excessive privileges, resulting in poorly implemented business controls over (e.g., financial) transactions.
	N There is a conflict of interest during user access management.	• Pressure from IT management on the security team under its authority results in excessive access rights been given to some users.
8 Software adoption and use	A Users fail to adopt new application software.	• Users refuse or are reluctant to use a new software application because of lack of training/communication, thereby foregoing efficiency gains.
	B Users use new software inefficiently.	• Users are provided with a new system for managing customer complaints that automates the workflow, but they keep emailing directly to the support persons in parallel.
	C Unintended use of new software applications appears.	• Business users are not using a new software application as intended, leading to productivity losses.
	D Enterprise users are unable to use the software to realize desired outcomes (e.g., required business model or organizational changes are not made).	• A new business application experiences an outage due to a lack of familiarity with a newly deployed server system.
	E Operational glitches arise when new software is made operational.	• The organization's public-facing website is intermittently inaccessible following the deployment of a software upgrade.

	Figure 6.2—Example Generic and Specific Risk Scenarios *(cont.)*		
	Risk Scenario Category	Example Generic Risk Scenarios	Example Specific Risk Scenarios
8	Software adoption and use *(cont.)*	F Critical application software malfunctions regularly.	• A critical customer-facing application becomes very unstable after the latest software change, sharply reducing the application's availability and causing numerous customer complaints.
		G Application software is obsolete (old technology, poorly documented, expensive to maintain, difficult to extend, not integrated in current architecture, etc.).	• A key business application is unstable due to older, unsupported technologies, resulting in periodic outages.
		H The enterprise is unable to revert to former versions of software when operational issues arise with the new version.	• A key business application is disabled due to a failure in the new version of the software, and the enterprise is unable to revert to the older version.
		I A software-induced corrupted data(base) leads to inaccessible data.	• A bug in the latest release of software results in corruption of data in a key business application, which makes the application unavailable.
		J Software-induced information integrity issues appear.	• A software change introduces new business and processing logic, but errors are causing information integrity problems, leading to wrong business decisions (e.g., credit approvals, customer order processing).
9	IT hardware	A New infrastructure is installed and, as a result, systems become unstable, leading to operational incidents (e.g., a bring your own device [BYOD] program).	• Outage of a key business application results from the instability of new Cisco network hardware.
		B Systems cannot handle transaction volumes when user volumes increase.	• A database is unable to handle transaction volumes, resulting in periodic outages. • An application is unable to handle transaction volumes, resulting in slow response times for customers.
		C Systems cannot handle system load when new applications or initiatives are deployed.	• Network throughput constraints introduce transaction delays when the new firewall is deployed.
		D Utilities (e.g., telecom, electricity) fail.	• Online consumer transactions are interrupted when a storm causes an electricity service disruption.
		E Hardware fails due to overheating and/or other environmental conditions, such as humidity.	• Online retail transactions are interrupted when the system hardware fails due to HVAC failure.
		F Hardware fails due to a lack of preventive maintenance.	• An HVAC system fails because of late maintenance, causing a computer room outage.
		G Hardware components are damaged, leading to (partial) destruction of data by internal staff.	• A disgruntled staff member physically destroys a hard drive, leading to data loss.
		H Portable media containing sensitive data (CDs, USB] drives, portable disks, etc.) are lost or disclosed.	• Customer information is compromised due to employees using unauthorized and unencrypted USB drives.

Figure 6.2—Example Generic and Specific Risk Scenarios (cont.)

Risk Scenario Category		Example Generic Risk Scenarios	Example Specific Risk Scenarios
9	IT hardware (cont.)	I	The enterprise experiences extended resolution time or support delays in cases of hardware incidents.	• Customers experience lengthy service delays due to outages of hardware that is no longer supported by the vendor.
		J	Hardware components are configured erroneously.	• After an upgrade of the storage system, several disks are incorrectly configured, making the applications using the data run erroneously.
		K	Hardware commissioning and decommissioning management is inadequate.	• A deficient hardware decommissioning procedure causes data to not be destroyed before the equipment is decommissioned and removed from the premises.
		L	Hardware configuration compliance management is inadequate.	• An operator's failure to comply with a hardware configuration procedure leaves a new piece of equipment poorly configured, causing it to be exploited and generating a service interruption.
		M	Vendor support comes to an end.	• The vendor of a legacy system used by the enterprise has ceased all support for the system, resulting in total service interruption in case of a hardware defect.
10	Internal and external security threats (hacker, malware, etc.)	A	Unauthorized (internal) users successfully break into systems.	• An internal user breaks into a key business application and deletes customer account information.
		B	Service interruption occurs due to a denial of service (DoS) attack.	• A distributed denial of service (DDoS) attack results in performance degradation of a key business website.
		C	A website is defaced.	• A flaw in the security of the main marketing website results in a website defacement that creates embarrassment for the enterprise.
		D	The enterprise experiences a malware attack.	• Critical operational servers are intruded by malware. • Laptops are regularly infection with malware. • The enterprise is forced to pay ransom due to a set of DoS attacks.
		E	Industrial espionage takes place.	• A foreign state or enterprise hacks the enterprise's systems and obtains important and confidential product and client information, reducing the enterprise's competitive position.
		F	Hacktivism occurs.	• A DDoS attack is perpetrated by opponents of the company's environmental policies, taking the website offline for more than 24 hours.
		G	A disgruntled employee implements a time bomb, which leads to data loss.	• Data in the central customer information database are deleted and made unrecoverable by a disgruntled employee who implements a data deletion time bomb.
		H	Company data are stolen through unauthorized access gained by a phishing attack.	• Employees use unauthorized USB devices, distributed on the parking lot, resulting in a loss of data. • An unaware employee opens a phishing email, resulting in unauthorized access to financial data.

Figure 6.2—Example Generic and Specific Risk Scenarios *(cont.)*		
Risk Scenario Category	**Example Generic Risk Scenarios**	**Example Specific Risk Scenarios**
10 **Internal and external security threats (hacker, malware, etc.)** *(cont.)*	I A foreign government issues attacks on critical systems.	• A foreign government attacks critical systems. • Organized crime groups intrude into systems for fraudulent transactions.
	J The data center is destroyed by staff (sabotage, etc.).	• Operations staff of the data center plant a bomb in the HVAC facilities.
	K A device with sensitive data is stolen.	• An authorized administrator steals a device with sensitive financial data.
	L IT equipment is accidentally damaged.	• The cleaning staff accidentally damage network cabinet equipment. • IT operations staff accidentally damage disk racks.
	M A key infrastructure component is stolen.	• A staff member of the data center cleaning crew steals the laptop managing the access control system to the data center, making it impossible to securely enter the data center.
	N Hardware (security devices, etc.) is intentionally tampered with.	• A technician tampers with the fingerprint recognition system for sensitive premises so the system allows anybody in.
11 **Third-party/ supplier incidents**	A An outsourcer performs inadequately in a large-scale, long-term outsourcing arrangement (e.g., through lack of supplier due diligence regarding financial viability, delivery capability and sustainability of supplier's service).	• Delivery of key product components is interrupted due to the financial failure of an outsourced provider.
	B The enterprise accepts unreasonable terms of business of IT suppliers (e.g., through lack of legal advice).	• Cost overruns occur due to poor negotiation of terms in agreements with IT service providers.
	C Vendors deliver inadequate support and services, not in line with the SLA.	• Delays in software fixes to the online retail website occur due to software vendors who fail to meet SLA requirements for timeliness and/or quality.
	D Noncompliance with software license agreements (use and/or distribution of unlicensed software, etc.) occurs.	• Legal action is brought by a software provider who identifies that the enterprise has violated the software license agreement by unmanaged distribution of the software.
	E The enterprise is unable to transfer to alternative suppliers due to overreliance or overdependence on the current supplier.	• Product delivery services are degraded due to overreliance on a single service provider that cannot scale up to meet increased demand.
	F IT services (especially cloud services) are purchased by the business without consultation/involvement of IT, resulting in an inability to integrate the service with in-house services.	• A cloud service provider is selected that is incompatible with internal applications.

Risk Scenario Category		Example Generic Risk Scenarios	Example Specific Risk Scenarios
Figure 6.2—Example Generic and Specific Risk Scenarios *(cont.)*			
11 **Third-party/ supplier incidents** *(cont.)*	G	Penalties are incurred due to noncompliance with SLAs.	• Legal action is brought against the enterprise for failure to meet contractual requirements for delivery of services.
	H	The SLA is inadequate to obtain required services.	• An inadequate SLA does not describe required service levels, which consequently are not delivered by the third-party provider, affecting staff productivity.
	I	Service is terminated.	• A service provider ceases to exist due to bankruptcy, leading to a service interruption.
	J	Operational service is interrupted by cloud providers.	• The cloud provider that hosts the enterprise's customer database has an operational malfunction of its systems, resulting in unavailability of the customer database for a day.
	K	Noncompliance with security requirements occurs.	• A cloud service provider is not compliant with contractual security requirements, which leads to a confidentiality breach with legal consequences for the enterprise.
	L	Third-party outsourcers do not deliver projects as per contractual agreements (any combination of exceeded budgets, quality problems, missing functionality, late delivery).	• The contractor for the new ERP system fails to deliver the new system by the promised deadline and will probably deliver one year late.
	M	Monitoring of the SLA with third-party providers is inadequate.	• Due to a lack of monitoring of the performance of a cloud provider, degradation of services and noncompliance with contractual service levels remain unnoticed, affecting staff productivity.
	N	The cloud provider loses data.	• The cloud provider that hosts the enterprise's customer database has a major malfunction of its systems, resulting in a loss of the previous month's customer updates.
12 **Noncompliance**	A	Noncompliance with local or international regulations (e.g., privacy, accounting, manufacturing, environmental) occurs.	• The enterprise experiences high fines and major damage to its brand due to unnecessary mistakes made against GDPR regulations. • Inadequate analysis of regulations leads to unintended noncompliance and, as a result, fines are incurred.
	B	The enterprise is unaware of potential regulatory changes that may have a business impact.	• The enterprise undertakes incomplete follow-up of changes in environmental regulations.
	C	The enterprise experiences operational obstacles caused by regulations.	• The regulator prevents cross-border data flow due to insufficient controls.
	D	The enterprise experiences compliance failures with internal procedures.	• Legal action is brought against the organization by personnel who cite a failure to comply with internal safety policies. • Users do not comply with internal policies on downloading external software, leading to infected systems and subsequent service interruptions.
	E	Cross-border regulations cause obstacles for the enterprise.	• The existence of different regulations between countries prohibits or inhibits information transfer, leading to additional costs and efficiency loss.

Risk Scenario Category			Example Generic Risk Scenarios	Example Specific Risk Scenarios
Figure 6.2—Example Generic and Specific Risk Scenarios *(cont.)*				
13	Geopolitical issues	A	The enterprise experiences service interruptions due to disruptive incidents (e.g., physical attack) in foreign premises.	• The data center in a foreign country is bombed and causes important supply chain problems.
		B	Government interference and national policies impact the enterprise.	• Government interference and national policies limit international service or product delivery.
		C	The enterprise experiences targeted action from government-sponsored groups or agencies.	• Targeted action against the enterprise results in destruction of critical infrastructure. • Targeted action from foreign entities obtains confidential product information, damaging the competitive position of the enterprise.
14	Industrial action	A	Facilities and buildings are not accessible because of a labor union strike.	• A reduction in productivity occurs because laborers at the production plant are unable to go to work due to a union strike.
		B	Third-party providers are not able to provide services because of a strike.	• External technicians cannot provide technical services because the premises are blocked by strikers, leading to service degradation.
		C	Key staff are not available as a result of external industrial action (e.g., a transportation or utilities strike).	• A reduction in productivity occurs because key staff are unable to go to work due to a transportation strike.
15	Acts of nature	A	An earthquake destroys or damages important IT infrastructure.	• An earthquake damages the IT operations center, resulting in the outage of online retail web services.
		B	A tsunami destroys critical premises.	• A tsunami destroys a major distribution center, resulting in an inability to meet product delivery deadlines.
		C	Major storms and tropical cyclones damage critical infrastructure.	• A major storm damages electrical infrastructure, resulting in the unavailability of grid-based power to the data center for more than two weeks.
		D	Major wildfires and/or flooding damage critical premises.	• A wildfire forces staff of a data center to abandon the premises, causing prolonged service interruptions. • Flooding of the premises causes irreparable damage to computer equipment.
		E	Changing environmental conditions affect the enterprise.	• A rising water table makes a critical location unusable, requiring relocation. • A rising temperature makes critical locations uneconomical to operate.
		F	A pandemic affects people and the economy.	• Staff are unable to access the office premises due to natural disaster (e.g., flood or COVID-19), causing efficiency degradation. • Permanent or prolonged loss of key employees during the pandemic causes service interruptions, knowledge gaps and efficiency degradation. • The enterprise is unable to acquire key resources (e.g., bandwidth) during the pandemic.

Figure 6.2—Example Generic and Specific Risk Scenarios *(cont.)*			
Risk Scenario Category		**Example Generic Risk Scenarios**	**Example Specific Risk Scenarios**
16	**Emerging technologies and innovation**	A New and important technology trends are not identified.	• A failure to leverage advancements in technology results in a loss of market share.
		B The enterprise fails to appreciate the value and potential (new functionality, process optimization) of new technologies.	• The enterprise incurs important opportunity costs by not adopting in a timely manner a new software platform for customer interactions.
		C The enterprise experiences problematic early adoption of new technology.	• Early adoption of new technology causes inconsistent stability and reliability, causing unanticipated interruptions to operations, demand for massive support and frequent updates, all resulting in efficiency degradation.
		D The enterprise fails to understand or address the risk linked to the adoption of new technology (e.g., lack of hardening guidance).	• Adoption of insufficiently hardened new technology may expose the enterprise to security risk and create service interruptions.
		E Emerging technology disappears from the market.	• New technology that has been adopted early disappears from the market, leading to service interruptions or additional cost for replacement technologies.
17	**Environmental**	A Equipment used by the enterprise is not environmentally friendly (e.g., power consumption, packaging).	• The use of herbicide sprayers out of accordance with established safety practices results in damage to endangered wildlife species.
		B Equipment used by the enterprise does not comply with environmental regulations or energy consumption standards.	• Computing equipment does not comply with environmental regulations, leading to penalties.
		C Equipment used by the enterprise does not comply with the latest energy consumption standards.	• The energy consumption by the enterprise's computing equipment is excessive and leads to higher cost.
18	**Data and information management**	A Sensitive information is discovered by unauthorized persons due to inefficient retaining/archiving/disposing of information.	• Accidental disclosure of sensitive client information due to failure to follow information handling guidelines (e.g., disposal of documents) leads to legal procedures.
		B Data are subject to intentional illicit or malicious modification.	• Intentional illicit or malicious modification of accounting data leads to an unfavorable audit opinion. • Intentional illicit or malicious modification of a patient's medical records leads to legal procedures.
		C Unauthorized disclosure of sensitive information through email or social media occurs.	• A design engineer shares new and confidential product information on his/her social media accounts with lots of followers. • An HR staff member shares confidential news on layoffs on his/her social media accounts, causing lots of press attention and major disruption in the organization.
		D The enterprise loses intellectual property (IP) and/or suffers leakage of competitive information.	• The enterprise loses IP and/or its competitive information is leaked due to key team members leaving the organization and taking information with them. • Hacking results in a loss of IP and/or leakage of competitive information.

	Figure 6.2—Example Generic and Specific Risk Scenarios *(cont.)*		
Risk Scenario Category		**Example Generic Risk Scenarios**	**Example Specific Risk Scenarios**
18 **Data and information management** *(cont.)*	E	The enterprise's information is of poor quality.	• Management of a service organization is unable to optimize its staff planning because of the poor quality of staff utilization and consistent communication.
	F	Master data management is inadequate.	• Due to poor master data management or poor data architecture, multiple instances of the same data exist on the enterprise's systems, leading to important customer service and efficiency degradation.
	G	Information classification is inadequate.	• Poor information classification leads to inadequate protection of information and disclosure of confidential marketing information, reducing the competitive advantage of the enterprise.

Chapter 7
Essentials of Risk Response

7.1 Risk Response Components

This chapter discusses the essential components of risk response. The topics discussed here include:

1. Risk avoidance
2. Risk mitigation
3. Risk sharing or transfer
4. Risk acceptance
5. Preliminary risk aggregation for response actions
6. Preliminary risk response selection and prioritization

The purpose of defining a risk response is to bring risk in line with the defined risk appetite for the enterprise after risk analysis. In other words, a response needs to be defined such that future residual risk (current risk with the risk response defined and implemented) is, as much as possible (usually depending on available budget), within risk tolerance limits. There may also be an exception process that allows some period of time to implement a risk response or for which management may decide to accept any risk, regardless of circumstances.

7.1.1 Risk Avoidance

Avoidance means exiting the activities or conditions that give rise to risk. Risk avoidance applies when no other risk response is adequate. This is the case when:

- There is no other cost-effective response that can succeed in reducing the impact of the realized risk below the defined thresholds for loss.
- The risk cannot be shared or transferred.
- The risk is deemed unacceptable by management

Some I&T-related examples of risk avoidance may include relocating a data center away from a region with significant natural hazards or declining to engage in a very large project when the business case shows a notable risk of failure.

7.1.2 Risk Mitigation

Risk mitigation means that once risk has been identified and analyzed, actions are taken to reduce the frequency and/or impact of a risk. The most common ways of mitigating risk include:

- Strengthening overall risk management practices. This can be assigning responsibility for risk identification and management of risk to those closest to the activities that generate significant risk. This type of activity also assists with raising awareness of risk across the enterprise.
- Embedding risk awareness activities into the regular business workflow so they become part of the regular course of daily activities. This allows staff to better understand and recognize risky behaviors before an incident materializes.
- Improving the risk management processes and developing relevant tolerances that cascade from the strategy all the way to the shop floor or front line of the enterprise

- Automating, where possible, the triggers or alerts that would identify or indicate when thresholds are out of tolerance
- Introducing or strengthening controls intended to reduce either the frequency or impact of a realized risk. Some techniques to do this are discussed in the following sections.

7.1.3 Risk Sharing or Transfer

Sharing means reducing risk frequency or impact by transferring or otherwise sharing a portion of the risk. Common techniques include insurance and outsourcing. Examples include taking out insurance coverage for IT-related (e.g., disaster recovery) or cyberincidents (e.g., data breach, ransomware), outsourcing part of the I&T activities, or sharing I&T project mitigations with the provider through fixed price arrangements or shared investment arrangements. In both a physical and legal sense these techniques do not relieve an enterprise of a risk but can involve the skills of another party in managing the risk and reduce the financial impact if an adverse event occurs. Risk ownership always stays with the enterprise, even when entering into a risk transfer, risk sharing or outsourcing arrangement.

During the risk analysis activities, it is important to be mindful that high-impact or imminent risk may need to be raised to a level of the organization for appropriate decision-making on response options. See section 1.1.1 for additional considerations for communicating and advising the appropriate accountable and responsible stakeholders in the enterprise. It may also be prudent to check assessment and analyses assumptions, techniques, and basic math or model calculations to ensure the analysis is accurate. Assumptions and techniques are equally important in qualitative and quantitative assessments. There is also an opportunity here to better understand the connection between PML/MFL and insurance coverage considerations for certain scenarios. It might be advisable to establish an insurance limit to cover the MFL and the annual premium or deductible around the PML.

During the risk analysis activities, it is important to be mindful that high-impact or imminent risk may need to be raised to a level of the organization for appropriate decision-making on response options.

7.1.4 Risk Acceptance

Risk acceptance means that no action is taken relative to a particular risk, and loss is accepted when/if it occurs. This is different from being ignorant of risk; accepting risk assumes that the risk is known, that is, an informed decision has been made by management to accept it as such. If an enterprise adopts a risk acceptance stance, it should carefully consider who can accept the risk—even more so with I&T risk. I&T risk should be accepted only by business management (and business process owners) in collaboration with and supported by the IT department or IT support function, and acceptance should be communicated to appropriate stakeholders, such as senior management and the board as necessary and dictated by policy. Identification or mitigation of every risk may not be relevant or cost effective. In addition, losses or incidents related to accepted risk should be tracked and communicated over time and accepted risk should be reassessed periodically for changes in the business landscape, assumptions or other factors.

7.1.5 Preliminary Risk Aggregation for Response Actions

Risk aggregation is the method or process by which individual risk areas may be joined together for the purpose of reporting, treatment, or to obtain an integrated risk profile or risk score. Decisions related to I&T risk management will be more beneficial to the enterprise if the risk is managed with a perspective of the end-to-end (business activity) aggregated view of all risk. An aggregated view of risk allows a complete and thorough review of risk appetite and risk tolerance, instead of having only a silo view of individual or partial risk items.

I&T risk is often grouped together by risk type or risk that would be addressed by a similar risk response or specific control treatment. For example, if there are repeated audit findings or control deficiencies in an organization's access management approach across many different business or mission areas, the organization may decide that an enterprise initiative in access management may resolve the issue.

There are different ways to perform risk aggregation. Considerations for aggregating risk and using risk maps when presenting risk information include:

- Ensure there is a uniform, consistent, agreed-on and communicated method for assessing frequency and impact of risk scenarios. The same method should be used to present aggregated risk. Using a consistent taxonomy for describing risk allows for aggregation and reporting on varying types of risk.

- Be cautious with the mathematics, and aggregate only the data and numbers that are meaningful. Do not aggregate data of different nature (e.g., on status of controls or operational IT metrics). Although separately these may be adequate risk indicators, they are meaningless when they are not associated with an ultimate business impact.

- Focus on risk for business activities and the most important indicators thereof and avoid focusing on adding up things that are easily measurable but less relevant. Reporting firewall attacks may be easy to measure, but if up-to-date security measures are in place, these attacks, although probably very frequent, may have little business impact.

- Do not aggregate risk information in such a way that it hides actionable detail. This may occur because of the organizational reporting level of responsibility issues that must be addressed by a certain organizational layer and must be visible to that layer, but may be aggregated and hidden from the next higher level of authority because no immediate action is required by that level. The root cause of risk must be visible to those responsible for managing it. Attention must be given to the aggregation algorithm that is used.

- Aggregation is possible in multiple dimensions (e.g., organizational units, types of risk items, business processes). The benefit of aggregation in business processes is that it reveals weak links to achieving successful business outcomes. Sometimes multiple views (using a combination of several dimensions) may be needed to satisfy risk management and business needs.

- Aggregate risk at the enterprise level, where risk can be considered in combination with all other risk the enterprise needs to manage (integration with ERM). Take into account the organizational structure (geographical split, business units, etc.) to set up a meaningful cascade of risk aggregation without losing sight of important specific risk items.

One benefit of aggregation is that the risk for the overall enterprise becomes highly visible and the need to define an enterprise response to the risk is more feasible or justifiable. Thus, aggregation allows the definition and implementation of a cost-efficient response to current risk and residual risk to be reduced within the defined risk appetite levels.

7.1.6 Preliminary Risk Response Selection and Prioritization

The previous sections listed the available risk response options. This section contains a brief discussion on the selection of an appropriate response (i.e., given the risk at hand), how to respond and how to choose among the available response options (**figure 7.1**). Any references to risk response should not be confused with incident or crisis responses. Incident response may be considered a response to a risk that has materialized and is causing the enterprise to experience consequences that are occurring and must be contained or acted upon before they become a crisis. Generally, the response to an incident is handled by an ad hoc or dedicated team with the requisite skill set to address the immediate impact. The difference between an incident and crisis is often one of a temporal nature; for example, an incident occurs and if some action is not taken quickly, the incident may result in a crisis situation. A crisis may also be related to the magnitude of the resulting loss from a realized risk; for example, if the impact of the loss was sufficient to cause an organization to cease business or suffer a major setback, that would be considered a crisis. The following parameters need to be taken into account during the risk response and prioritization process:

- Cost of the response (e.g., in the case of risk transfer, the cost of the insurance premium; in the case of risk mitigation, the cost of implementing, maintaining and testing the controls)

- The importance of the risk addressed by the response (i.e., its priority or rank on the risk register)

- The enterprise's capability to implement and maintain the response over time. The more mature an enterprise is in its risk management capability, the better the responses that can be implemented. When the enterprise is rather immature, some very basic responses may be used and improved over time.

- The effectiveness of the response (i.e., the extent to which the response activities will reduce the frequency or impact of the risk if it should be realized)

- Other I&T-related investments (the ever-present competition between investing in risk response measures and other I&T investments)

- Other responses (one response may address several risks while another may not), as in the case of risk that can be aggregated and addressed with a common response

Sometimes the comprehensive required effort or resources for risk responses (e.g., the collection of controls that need to be implemented or strengthened) will exceed available capability of the enterprise. In this case, decisions on prioritization and organizational skill and expertise are required. Possible risk response options may be grouped as follows:

- Quick wins, which are very short-term, time-efficient and effective responses on high-impact risk

- Compliance obligations for which there is a nonnegotiable requirement. Managing the risk of noncompliance should be done in conjunction with other risk responses to avoid duplicate or overlapping work.

- Business case to be made, with more expensive or difficult responses to high-impact risk requiring careful analysis and management decision on investments. Responses in this category may also include decisions to outsource the management of risk that the organization has no capability to address appropriately.

- Deferring and/or monitoring conditions to determine if changes to the identified risk or the environment would warrant a different response

Figure 7.1—Risk Response Options and Prioritization

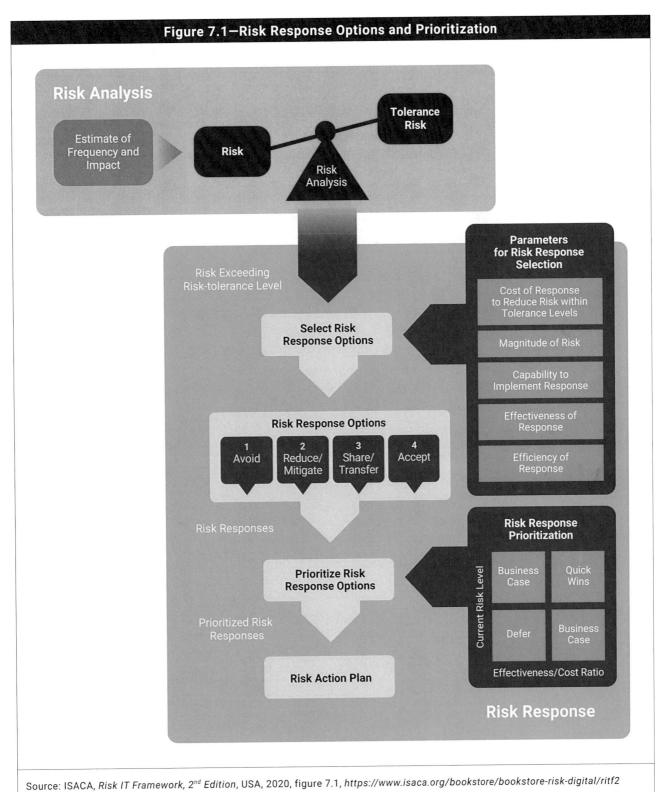

Source: ISACA, *Risk IT Framework, 2nd Edition*, USA, 2020, figure 7.1, *https://www.isaca.org/bookstore/bookstore-risk-digital/ritf2*

The following are aspects of prioritization criteria that merit consideration when determining which risk to address first, second, third and so on:

- Focus on the organization's mission and strategic objectives as the starting point to determine which risk, if realized, would have the greatest impact on the enterprise. Brainstorm scenarios until there is a robust set of things that are plausible and relevant to the enterprise, mission, level of response capability and dependence on third parties.

- Define the enterprise's most important products and services and the underlying technology that supports the delivery of those products or services. Looking at a technology asset in the context of the mission and strategy helps determine the criticality of the asset. The criticality of an asset is a key criterion to use in determining which risk to address first.

- From the list of enterprise products and services, make a list of vendors, service providers, suppliers or third parties that provide some or all of the resources needed to deliver each product or service. Looking at service providers and vendors in the context of mission and strategy helps determine the criticality of third parties and subsequently helps prioritize the risk.

- Perform research and ask subject matter experts inside and outside the organization to help in determining the probability, or likelihood, of a certain scenario materializing. Some threats are not relevant to the organization or the organization may not have the vulnerability that allows the threat to materialize.

- Focus on the risk with the largest potential impact on the enterprise. Which risk, if realized, would have the greatest impact on the ability of the organization to continue to deliver its products and services?

- Ask the senior leaders or board of directors to vote on which scenarios potentially would have the greatest impact on the enterprise. Their answers may be surprising and may also provide great insights to help refine the risk appetite and risk tolerance statements.

- Assess the enterprise's capability to detect and respond to a given scenario or set of scenarios. Some organizations fall below the cybersecurity poverty line[19] and are not able to adequately to respond to one or more risk scenarios. This is a real risk to many organizations, especially those who continue to operate in deep technical debt, and the enterprise senior leadership needs to be aware of this to best decide a course of action.

The main drivers for risk management include the need to improve decision-making in enterprises, align risk management resources to address the risk with the greatest potential impact on the enterprise, and build the capabilities and gather the resources needed to detect and respond to realized risk.

---

[19] The [cyber]security poverty line is a term coined by Wendy Nather in 2013; see Nather, W.; "The Longevity Challenge in Infosec," 4 October 2016, RSA Conference, *https://www.rsaconference.com/industry-topics/blog/the-longevity-challenge-in-infosec.*

APPENDIX A
Risk Resources

This appendix provides resources that are available in the public domain or from an international standards organization to aid in the risk management journey. Publications that are proprietary to an individual economic sector, such as finance, are excluded from this publication.

There are several risk taxonomies available that may be informative:

- A Taxonomy of Threats for Complex Risk Management[20]
- A Taxonomy of Operational Cyber Security Risks Version 2[21]
- Open Risk Taxonomy[22]
- OpenFAIR Risk Taxonomy[23]

Examples of standards and frameworks that may be useful sources of good practices include:

- *Risk IT Framework*, ISACA[24]
- COSO *Enterprise Risk Management—Integrated Framework*[25]
- *Operationally Critical Threat, Asset and Vulnerability Evaluation (OCTAVE) Framework* Allegro[26]
- ISO/IEC 27005:2011 *Information technology — Security techniques — Information security risk management*[27]
- ISO 31000:2009 *Risk management — Principles and guidelines*[28]
- IEC 31010:2009 *Risk management — Risk assessment techniques*[29]
- NIST Special Publication (SP) 800-30 Revision 1, *Guide for Conducting Risk Assessments*[30]
- NIST Special Publication (SP) 800-39 *Managing Information Security Risk: Organization, Mission, and Information System View*[31]

[20] University of Cambridge, Judge Business School, "A Taxonomy of Threats for Complex Risk Management," *https://www.jbs.cam.ac.uk/faculty-research/centres/risk/publications/managing-multi-threat/a-taxonomy-of-threats-for-complex-risk-management/*

[21] Cebula, J.; Popeck, M.; Young, L.; "A Taxonomy of Operational Cyber Security Risks Version 2," CMU/SEI-2014-TN-006, Software Engineering Institute, Carnegie Mellon University, 2014, *http://resources.sei.cmu.edu/library/asset-view.cfm?AssetID=91013*

[22] Papadopoulos, P.; "Open Risk Taxonomy," OpenRisk, 24 June 2015

[23] The Open Group, "Risk Taxonomy (O-RT), Version 2.0," 2009, *https://publications.opengroup.org/c13k*

[24] ISACA, *Risk IT Framework, 2nd Edition*, 2020, USA, *www.isaca.org/bookstore/bookstore-risk-digital/ritf2*

[25] Committee of Sponsoring Organizations (COSO), *Enterprise Risk Management—Integrated Framework*, June 2017, *https://www.coso.org/Pages/erm.aspx*

[26] Alberts, C.J.; Behrens, S.; Pethia, R.D.; Wilson, W.R; *Operationally Critical Threat, Asset, and Vulnerability Evaluation (OCTAVE) Framework*, Version 1.0, Carnegie Mellon University Software Engineering Institute, September 1999, *https://resources.sei.cmu.edu/asset_files/TechnicalReport/1999_005_001_16769.pdf*

[27] International Organization for Standardization (ISO®), ISO/IEC 27005:2011, *Information technology — Security techniques — Information security risk management*, June 2011, *https://www.iso.org/standard/56742.html*

[28] ISO, ISO 31000:2009, *Risk management — Principles and guidelines*, November 2009, *https://www.iso.org/standard/43170.html*

[29] ISO, IEC 31010:2009, *Risk management — Risk assessment techniques*, November 2009, *https://www.iso.org/standard/51073.html*

[30] NIST, SP 800-30 Rev. 1, *Guide for Conducting Risk Assessments*, September 2012, *https://csrc.nist.gov/publications/detail/sp/800-30/rev-1/final*

[31] NIST, SP 800-39, *Managing Information Security Risk: Organization, Mission, and Information System View*, March 2011, *https://csrc.nist.gov/publications/detail/sp/800-39/final*

Page intentionally left blank

APPENDIX B
Glossary

TERM	DEFINITION
Business continuity	Preventing, mitigating and recovering from disruption **Scope Notes:** The terms 'business resumption planning', 'disaster recovery planning' and 'contingency planning' also may be used in this context; they focus on recovery aspects of continuity, and for that reason the 'resilience' aspect should also be taken into account. COBIT 5 and COBIT 2019 perspective
Business goal	The translation of the enterprise's mission from a statement of intention into performance targets and results.
Business impact analysis (BIA)	A process to determine the impact of losing the support of any resource. **Scope Notes:** The BIA assessment study will establish the escalation of that loss over time. It is predicated on the fact that senior management, when provided reliable data to document the potential impact of a lost resource, can make the appropriate decision.
Business impact analysis/assessment (BIA)	Evaluating the criticality and sensitivity of information assets. An exercise that determines the impact of losing the support of any resource to an enterprise, establishes the escalation of that loss over time, identifies the minimum resources needed to recover, and prioritizes the recovery of processes and the supporting system. **Scope Notes:** This process also includes addressing: • Income loss • Unexpected expense • Legal issues (regulatory compliance or contractual) • Interdependent processes • Loss of public reputation or public confidence
Business impact	The net effect, positive or negative, on the achievement of business objectives.
Business interruption	Any event, whether anticipated (i.e., public service strike) or unanticipated (i.e., blackout) that disrupts the normal course of business operations at an enterprise.
Business objective	A further development of the business goals into tactical targets and desired results and outcomes.
Business process control	The policies, procedures, practices and organizational structures designed to provide reasonable assurance that a business process will achieve its objectives. **Scope Notes:** COBIT 5 and COBIT 2019 perspective

TERM	DEFINITION
Business risk	A probable situation with uncertain frequency and magnitude of loss (or gain).
Compensating control	An internal control that reduces the risk of an existing or potential control weakness resulting in errors and omissions.
Control framework	A set of fundamental controls that facilitates the discharge of business process owner responsibilities to prevent financial or information loss in an enterprise.
Control objective	A statement of the desired result or purpose to be achieved by implementing control procedures in a particular process.
Control practice	Key control mechanism that supports the achievement of control objectives through responsible use of resources, appropriate management of risk and alignment of IT with business.
Control risk	The risk that a material error exists that would not be prevented or detected on a timely basis by the system of internal controls (See Inherent risk).
Control risk self-assessment	A method/process by which management and staff of all levels collectively identify and evaluate risk and controls with their business areas. This may be under the guidance of a facilitator such as an auditor or risk manager.
Control weakness	A deficiency in the design or operation of a control procedure. Control weaknesses can potentially result in risk relevant to the area of activity not being reduced to an acceptable level (relevant risk threatens achievement of the objectives relevant to the area of activity being examined). Control weaknesses can be material when the design or operation of one or more control procedures does not reduce to a relatively low level the risk that misstatements caused by illegal acts or irregularities may occur and not be detected by the related control procedures.
Corporate governance	The system by which enterprises are directed and controlled. The board of directors is responsible for the governance of their enterprise. It consists of the leadership and organizational structures and processes that ensure the enterprise sustains and extends strategies and objectives.
Corrective control	Designed to correct errors, omissions and unauthorized uses and intrusions, once they are detected.
Countermeasure	Any process that directly reduces a threat or vulnerability.
Control	The means of managing risk, including policies, procedures, guidelines, practices or organizational structures, which can be of an administrative, technical, management, or legal nature. **Scope Notes:** Also used as a synonym for safeguard or countermeasure. See also Internal control.
Criticality	The importance of a particular asset or function to the enterprise, and the impact if that asset or function is not available
Culture	A pattern of behaviors, beliefs, assumptions, attitudes and ways of doing things **Scope Notes:** COBIT 5 and COBIT 2019 perspective

TERM	DEFINITION
Detailed IS controls	Controls over the acquisition, implementation, delivery and support of IS systems and services made up of application controls plus those general controls not included in pervasive controls.
Enterprise risk management (ERM)	The discipline by which an enterprise in any industry assesses, controls, exploits, finances and monitors risk from all sources for the purpose of increasing the enterprise's short- and long-term value to its stakeholders.
Event type	For the purpose of IT risk management, one of three possible sorts of events: threat event, loss event and vulnerability event. **Scope Notes:** Being able to consistently and effectively differentiate the different types of events that contribute to risk is a critical element in developing good risk-related metrics and well-informed decisions. Unless these categorical differences are recognized and applied, any resulting metrics lose meaning and, as a result, decisions based on those metrics are far more likely to be flawed.
Event	Something that happens at a specific place and/or time
Exposure	The potential loss to an area due to the occurrence of an adverse event.
Framework	**Scope Notes:** See control framework and IT governance framework.
Frequency	A measure of the rate by which events occur over a certain period of time
Impact	Magnitude of loss resulting from a threat exploiting a vulnerability
Information technology (IT)	The hardware, software, communication and other facilities used to input, store, process, transmit and output data in whatever form.
Inherent risk	The risk level or exposure without taking into account the actions that management has taken or might take (e.g., implementing controls)
Intangible asset	An asset that is not physical in nature. **Scope Notes:** Examples include: intellectual property (patents, trademarks, copyrights, processes), goodwill, and brand recognition
Internal controls	The policies, procedures, practices and organizational structures designed to provide reasonable assurance that business objectives will be achieved and undesired events will be prevented or detected and corrected.
IT incident	Any event that is not part of the ordinary operation of a service that causes, or may cause, an interruption to, or a reduction in, the quality of that service.
IT risk issue	1. An instance of IT risk. 2. A combination of control, value and threat conditions that impose a noteworthy level of IT risk.
IT risk profile	A description of the overall (identified) IT risk to which the enterprise is exposed.
IT risk register	A repository of the key attributes of potential and known IT risk issues. Attributes may include name, description, owner, expected/actual frequency, potential/actual magnitude, potential/actual business impact, disposition.

TERM	DEFINITION
IT risk scenario	The description of an IT-related event that can lead to a business impact.
IT risk	The business risk associated with the use, ownership, operation, involvement, influence and adoption of IT within an enterprise.
IT-related incident	An IT-related event that causes an operational, developmental and/or strategic business impact.
Key risk indicator (KRI)	A subset of risk indicators that are highly relevant and possess a high probability of predicting or indicating important risk. **Scope Notes:** See also Risk Indicator.
Likelihood	The probability of something happening
Loss event	Any event during which a threat event results in loss. **Scope Notes:** From Jones, J.; "FAIR Taxonomy," Risk Management Insight, USA, 2008
Pervasive IS control	General control designed to manage and monitor the IS environment and which, therefore, affects all IS-related activities.
Preventive control	An internal control that is used to avoid undesirable events, errors and other occurrences that an enterprise has determined could have a negative material effect on a process or end product.
Process	Generally, a collection of activities influenced by the enterprise's policies and procedures that takes inputs from a number of sources, (including other processes), manipulates the inputs and produces outputs. **Scope Notes:** Processes have clear business reasons for existing, accountable owners, clear roles and responsibilities around the execution of the process, and the means to measure performance.
Reputation risk	The current and prospective effect on earnings and capital arising from negative public opinion. **Scope Notes:** Reputation risk affects a bank's ability to establish new relationships or services, or to continue servicing existing relationships. It may expose the bank to litigation, financial loss or a decline in its customer base. A bank's reputation can be damaged by Internet banking services that are executed poorly or otherwise alienate customers and the public. An Internet bank has a greater reputation risk as compared to a traditional brick-and-mortar bank, because it is easier for its customers to leave and go to a different Internet bank and since it cannot discuss any problems in person with the customer.
Residual risk	The remaining risk after management has implemented a risk response.
Risk	The combination of the likelihood of an event and its impact. (ISO/IEC 73).

TERM	DEFINITION
Risk acceptance	If the risk is within the enterprise's risk tolerance or if the cost of otherwise mitigating the risk is higher than the potential loss, the enterprise can assume the risk and absorb any losses
Risk aggregation	The process of integrating risk assessments at a corporate level to obtain a complete view on the overall risk for the enterprise.
Risk analysis	1. A process by which frequency and magnitude of IT risk scenarios are estimated. 2. The initial steps of risk management: analyzing the value of assets to the business, identifying threats to those assets and evaluating how vulnerable each asset is to those threats. **Scope Notes:** It often involves an evaluation of the probable frequency of a particular event, as well as the probable impact of that event.
Risk appetite	The amount of risk, on a broad level, that an entity is willing to accept in pursuit of its mission.
Risk assessment	A process used to identify and evaluate risk and its potential effects. **Scope Notes:** Risk assessments are used to identify those items or areas that present the highest risk, vulnerability or exposure to the enterprise for inclusion in the IS annual audit plan. Risk assessments are also used to manage the project delivery and project benefit risk.
Risk avoidance	The process for systematically avoiding risk, constituting one approach to managing risk
Risk culture	The set of shared values and beliefs that governs attitudes toward risk-taking, care and integrity, and determines how openly risk and losses are reported and discussed.
Risk evaluation	The process of comparing the estimated risk against given risk criteria to determine the significance of the risk. [ISO/IEC Guide 73:2002].
Risk factor	A condition that can influence the frequency and/or magnitude and, ultimately, the business impact of IT-related events/scenarios
Risk indicator	A metric capable of showing that the enterprise is subject to, or has a high probability of being subject to, a risk that exceeds the defined risk appetite
Risk management	1. The coordinated activities to direct and control an enterprise with regard to risk **Scope Notes:** In the International Standard, the term "control" is used as a synonym for "measure." (ISO/IEC Guide 73:2002) 2. One of the governance objectives. Entails recognizing risk; assessing the impact and likelihood of that risk; and developing strategies, such as avoiding the risk, reducing the negative effect of the risk and/or transferring the risk, to manage it within the context of the enterprise's risk appetite. **Scope Notes:** COBIT 5 perspective

TERM	DEFINITION
Risk map	A (graphic) tool for ranking and displaying risk by defined ranges for frequency and magnitude.
Risk mitigation	The management of risk through the use of countermeasures and controls
Risk portfolio view	1. A method to identify interdependencies and interconnections among risk, as well as the effect of risk responses on multiple types of risk. 2. A method to estimate the aggregate impact of multiple types of risk (e.g., cascading and coincidental threat types/scenarios, risk concentration/correlation across silos) and the potential effect of risk response across multiple types of risk.
Risk owner	The person in whom the organization has invested the authority and accountability for making risk-based decisions and who owns the loss associated with a realized risk scenario. **Scope Notes:** The risk owner may not be responsible for the implementation of risk treatment.
Risk reduction	The implementation of controls or countermeasures to reduce the likelihood or impact of a risk to a level within the organization's risk tolerance.
Risk response	Risk avoidance, risk acceptance, risk sharing/transfer, risk mitigation, leading to a situation that as much future residual risk (current risk with the risk response defined and implemented) as possible (usually depending on budgets available) falls within risk appetite limits.
Risk scenario	The tangible and assessable representation of risk. **Scope Notes:** One of the key information items needed to identify, analyze and respond to risk (COBIT 2019 objective APO12)
Risk sharing	**Scope Notes:** See risk transfer
Risk statement	A description of the current conditions that may lead to the loss; and a description of the loss Source: Software Engineering Institute (SEI) **Scope Notes:** For a risk to be understandable, it must be expressed clearly. Such a treatment must include a description of the current conditions that may lead to the loss; and a description of the loss.
Risk tolerance	The acceptable level of variation that management is willing to allow for any particular risk as the enterprise pursues its objectives.
Risk transfer	The process of assigning risk to another enterprise, usually through the purchase of an insurance policy or by outsourcing the service. **Scope Notes:** Also known as risk sharing
Risk treatment	The process of selection and implementation of measures to modify risk (ISO/IEC Guide 73:2002).
Safeguard	Safeguard A practice, procedure or mechanism that reduces risk.

TERM	DEFINITION
Security incident	A series of unexpected events that involves an attack or series of attacks (compromise and/or breach of security) at one or more sites. A security incident normally includes an estimation of its level of impact. A limited number of impact levels are defined and, for each, the specific actions required and the people who need to be notified are identified.
Security/transaction risk	The current and prospective risk to earnings and capital arising from fraud, error and the inability to deliver products or services, maintain a competitive position, and manage information. **Scope Notes:** Security risk is evident in each product and service offered, and it encompasses product development and delivery, transaction processing, systems development, computing systems, complexity of products and services and the internal control environment. A high level of security risk may exist with Internet banking products, particularly if those lines of business are not adequately planned, implemented and monitored.
Threat agent	Methods and things used to exploit a vulnerability. **Scope Notes:** Examples include determination, capability, motive and resources.
Threat analysis	An evaluation of the type, scope and nature of events or actions that can result in adverse consequences; identification of the threats that exist against enterprise assets **Scope Notes:** The threat analysis usually defines the level of threat and the likelihood of it materializing.
Threat event	Any event during which a threat element/actor acts against an asset in a manner that has the potential to directly result in harm.
Threat vector	The path or route used by the adversary to gain access to the target
Threat	Anything (e.g., object, substance, human) that is capable of acting against an asset in a manner that can result in harm. **Scope Notes:** A potential cause of an unwanted incident (ISO/IEC 13335)
Vulnerability analysis	A process of identifying and classifying vulnerabilities.
Vulnerability event	Any event during which a material increase in vulnerability results. Note that this increase in vulnerability can result from changes in control conditions or from changes in threat capability/force. **Scope Notes:** From Jones, J.; "FAIR Taxonomy," Risk Management Insight, USA, 2008
Vulnerability scanning	An automated process to proactively identify security weaknesses in a network or individual system
Vulnerability	A weakness in the design, implementation, operation or internal control of a process that could expose the system to adverse threats from threat events
Zero-day-exploit	A vulnerability that is exploited before the software creator/vendor is even aware of its existence

Page intentionally left blank